THE CAMBRIDGE LIBRARY OF
MODERN SCIENCE

GENERAL EDITOR: C. P. SNOW

BACKGROUND TO MODERN SCIENCE

BACKGROUND TO MODERN SCIENCE

*Ten Lectures at Cambridge arranged by
the History of Science Committee*
1936

by

F. M. CORNFORD

SIR W. DAMPIER

LORD RUTHERFORD

W. L. BRAGG

F. W. ASTON

SIR A. S. EDDINGTON

J. A. RYLE

G. H. F. NUTTALL

R. C. PUNNETT

J. B. S. HALDANE

edited by
JOSEPH NEEDHAM
and
WALTER PAGEL

CAMBRIDGE
at the University Press
1938

CAMBRIDGE
UNIVERSITY PRESS

University Printing House, Cambridge CB2 8BS, United Kingdom

Cambridge University Press is part of the University of Cambridge.

It furthers the University's mission by disseminating knowledge in the pursuit of education, learning and research at the highest international levels of excellence.

www.cambridge.org
Information on this title: www.cambridge.org/9781107495005

© Cambridge University Press 1938

First published 1938
First paperback edition 2015

A catalogue record for this publication is available from the British Library

ISBN 978-1-107-49500-5 Paperback

CONTENTS

INTRODUCTION

In the understanding of man's ascent from primitive savagery to the relatively high level of civilization he has now attained, no study is more important than the history of science and technology. Through science man has reached a reliable knowledge of the properties of the world in which he finds himself; through applied science, or technique, he has succeeded in making himself ever more independent of his environment. There is a clear continuity here with the prior process of biological evolution itself, and biologists have long been ready to see in the studies of the sociologist the logical extrapolation, on a higher plane, of phenomena already familiar to them. The historian, on the other hand, has been more reluctant to consider these phenomena. Preoccupation with political, constitutional, and dynastic changes left him with little interest for the lines of thought of the scientific investigators, the successes of the technical improvers of trades, or the life of the common people in which these effects, so obscure, and yet so fundamental, worked themselves out. The rise of economic history, however, has created within the realm of historical studies an altogether different atmosphere, and to-day we may find everywhere substantial agreement on the importance of research and teaching in the history of science and technology.

From this *rapprochement* between historians proper, and historians of science, much may be expected. If in former times historians concerned themselves only with military or political history, it was no doubt because of that very ancient separation between scholarly administrator or warrior on the one hand and manual worker on the other; a separation in which all social prestige was accorded to the former groups, and the affairs of the latter were not thought worthy of intellectual consideration. Yet the entire structure of modern science which has arisen since the Renaissance has depended, and still depends, upon the union of intellectual activity and manual work. It is not perhaps surprising, moreover, that in these circumstances a kind of inverted "snobism" should have arisen in the sciences. Thus a very eminent biologist was once heard to remark that the history of science was a study fit only for those who, by reason of age or infirmity, were incapable of making serious laboratory experiments; no sane person in the prime of life, he thought, would be likely to show any interest in it. Deprived in this way of the support of historians on the one hand and of scientists on the other, historians of science have tended too much to fall into mere antiquarianism, and to suppose that their theme can adequately be treated by the stringing together of a series of individual biographies. From these defects the history of science may henceforward be free, if it will but accept the support now willingly offered by historians and scientists alike and go forward to present the history of scientific thought always in relation to the social and economic background of the time.

In Cambridge, in the University of Harvey and

Newton, of Whewell and Darwin, the great development of scientific research and teaching during the latter part of the nineteenth and throughout the present century, led to a certain forgetfulness of the value of the history of science as an element in the history of civilization. During the last ten years, however, the belief has been steadily growing that the lack of any organization for the study of the history of science in Cambridge constituted a serious gap. Two efforts were made in 1936 towards remedying this situation. In the summer, largely owing to the labours of Dr H. H. Thomas, pieces of scientific apparatus of historic interest were collected from the various laboratories and colleges, and exhibited for a week in one of the rooms of the Old Schools, then newly reopened. The exhibition aroused wide interest outside the University, and although it has so far proved impossible to keep the collection together, following the admirable example of the Old Ashmolean Museum at Oxford, the catalogue made for it now gives at least a knowledge of the whereabouts of these irreplaceable objects. It is greatly to be wished that a permanent home for the collection could be found.

The second effort consisted of the starting of a committee, with the blessing of the University authorities, by the two Faculty Boards of Biology and the Faculty Board of Physics and Chemistry, for arranging courses of lectures on the history of science. The committee now consists of Sir William Dampier, Dr H. H. Thomas, Prof. Green, Mr Bernal, Mr Butterfield, Prof. Postan, Prof. Harris, Prof. Webb, Prof. Hutton, Mr White, Dr Needham (chairman) and Mr Pagel (secretary). The first act of the committee was to arrange for an inaugural course of lectures on the modern period, 1895–1935, by scientific investigators who had themselves made fundamental contributions to science during that time. We now have the honour of reproducing these lectures in the present book.

In subsequent terms, specific historical periods of an earlier date were discussed, and the attendance of students (undergraduates of every year and every faculty, research workers, students taking advanced classes in science, etc.) overwhelmingly large during the first term, diminished to more manageable proportions, and keeps at a high level. The Lent term, 1937, was devoted to a "survey of the history of science" and included "Chemistry before Boyle" (Mr Bernal), "Biology from Galen to Harvey" (Dr Needham), "Astronomy 1700–1800" (Sir A. S. Eddington), "Biological Theory from Linnaeus to Darwin" (Dr Thomas), "The Chemical Revolution in the Eighteenth Century" (Mr McKie), and "Franklin and Faraday" (Prof. Ferguson). The year 1937–38 opened with an introductory lecture by Dr Charles Singer, and the Michaelmas term was given to science before 1500, while the Lent term dealt with science between 1500 and 1800. Among the lectures in the former series were "Science in the Ancient Empires" (Prof. Gordon Childe), "Greek Astronomy and Mathematics" (Sir Thomas Heath), "Greek Medicine" (Dr Jones), "Early Alchemy" (Prof. Read), "The Science of Islam" (Dr Holmyard), "The Relation of Science to Industry before 1500" (Prof. Postan). The Lent term, 1938, began with "Science and the Arts of the Renaissance" (Prof. Saxl) and went on to "Renaissance Botany" (Dr Arber), "Biology from Paracelsus to Malpighi" (Mr Pagel), "Descartes and Newton" (Prof. Andrade), "Science and the Philosophy of Matter in the Eighteenth Century" (Miss MacDonald), "Biology and the Social Background in the Seventeenth Century" (Dr Needham), and "The Relation of Science to Industry after 1500" (Mr Pledge).

It will be seen from these lists that the committee has been fortunate in being able not only to draw upon considerable stores of learning already present in Cambridge, but also to prevail upon scholars, not resident there, to come and address this keen and enthusiastic audience. In the long run there ought, of course, to be really well-organized facilities for the study of the history of science and technology in Cambridge, facilities which could hardly be obtained without the foundation of a chair in the subject and a small department for the accommodation of a few research workers. But doubtless this subject, like so many others, will have to wait for the endowment of some wealthy benefactor. In the meantime the committee believes that its lectures are meeting a real need, and by the co-operation of a large number of lecturers, the student has access to the largest stores of learning at the minimum expense.

In conclusion, the Editors would like to take this opportunity of thanking all those who by their valuable aid have helped to establish these new beginnings in the study of the history of science in Cambridge. In particular, the distinguished lecturers whose contributions are here printed deserve the warmest thanks. It was for us a moving experience to see the great concourse of students, many having to stand or to sit on the floor, which gathered to hear these expositions of progress in the sciences during the past forty years by those who had themselves taken some of the foremost parts in it. Nor can we forget that we are here privileged to print the last public lectures which two of our lecturers (Lord Rutherford[1] and Prof. Nuttall) were to give. It is our hope that all these efforts will end by the general recognition of the history of science and technology, both in Cambridge and elsewhere, as the great cultural subject which it is.

J. N.
W. P.

August 1938

---

[1] The Editors are deeply indebted to Mr J. A. Ratcliffe, who undertook the preparation of Lord Rutherford's lectures for press after his death.

I. GREEK NATURAL PHILOSOPHY AND MODERN SCIENCE

by

F. M. CORNFORD
Laurence Professor of Ancient Philosophy,
Cambridge

GREEK NATURAL PHILOSOPHY
AND MODERN SCIENCE

WHEN I was invited to contribute to this course a single lecture on Greek Philosophy and Science, my first impulse was to reply: Greek philosophy began when Thales of Miletus successfully predicted an eclipse of the sun in 585 B.C. and ended in A.D. 529 when the Christian Emperor Justinian closed the schools of Athens. What can I say, in fifty minutes, about a development of thought which covered eleven centuries—a longer span than separates ourselves from the reign of King Alfred?

Plainly, I must limit myself to a few general considerations. Of these the most relevant will be the differences distinguishing the Greek study of Nature from the natural science of our modern period. These differences are obvious on the surface, but the underlying reasons for them may easily be overlooked. The man of science to-day works at his own field within the horizon of a certain outlook, and using a certain apparatus of concepts which are the common property of his contemporaries. If he is not a philosopher or a psychologist, he may tacitly assume that this outlook and this apparatus are the only possible ones and have always been common property, imposed upon any student of Nature by Nature herself. He may then be puzzled, if he should dip into the pages of Aristotle or Lucretius. He may there light upon some startling anticipations of recent discoveries, but he will find them embedded in a mass of what looks to him like nonsense. He may conclude that the ancients were like clever children, with some bright ideas, but with untrained minds, who had advanced only a little way along the one right approach to truth.

This illusion is rampant in histories of philosophy and science. It is fostered by writers on the classics who catch at every chance of showing that the past they love is not out of date. But one purpose of this course of lectures should be to point out that the ancients were not moderns in the stage of infancy or adolescence. The Graeco-Roman culture was a self-contained growth, with its own infancy, adolescence, maturity, and decay. After the Dark Age and the Middle Age, the modern science of Nature starts at the Renaissance with a fresh motive impulse. The questions it asks are different questions. Its method is a new method, dictated by the need to meet those new questions with an appropriate answer.

You know better than I do what you are trying to find out here in your laboratories, and how and why you go about your task. I am told that you proceed by a method of tentative hypotheses, suggested by careful observation of facts, and controlled by no less careful experiment. Your objective has been described (at least till very recently) as the discovery of laws of cause and effect, invariable sequences of phenomena. And your motive—what is your motive? Shall we say: a pure and dispassionate love of truth for its own sake? I will accept that answer gladly; long may it remain as true as it is now in Cambridge. But there are some people who think that truth is the same thing as usefulness, and that the study of Nature really aims at the control of natural forces as a means to a further end. Some, again, would define that end as the increase of wealth and material comfort, and increase of power, which may itself be used to destroy, not only the comfort, but the lives, of our competitors in the scramble for wealth. Hence the subsidies lavished on natural science by War Departments and captains of industry. Hence the unabashed emergence of Nordic physics in central Europe and of proletarian physics farther East. Your very protons and electrons are suspected of capitalist or Marxian sympathies. Your neutrons are not to be politically neutral.

Now if that is a roughly true picture of natural science in the last four centuries, it differs in every respect—in method, in objective, and in underlying impulse—from the physical speculation of antiquity. My purpose is to bring out these differences and to raise, if I cannot wholly answer, the question why they exist.

First, let me indicate the limits of my subject. Other speakers are to deal with Greek mathematics and biology. "Science and Philosophy" in the title of this lecture must be taken to mean what the Greeks called Physics or "the inquiry into the nature of things". In this field all the most important and original work was done in the three centuries from 600 to 300 B.C. After Aristotle's death in 322 physics fell into the background; philosophers became preoccupied with the quest of a moral or religious faith that would make human life bearable.

In those first three centuries no line was drawn distinguishing philosophy from the study of Nature. Before Aristotle there were no separate branches of natural science. The word for science (knowledge, ἐπιστήμη) was applied rather to mathematics, because mathematics deals with exactly defined unchanging objects and demonstrable truths, and so could claim to yield knowledge in the fullest sense. Physics was known as "the inquiry into the nature of things". We should speak of it rather by its older name, natural philosophy. Accordingly, we are now concerned only with the natural philosophy of the period ending with the school of Aristotle.

Let us begin with method and procedure. In this period, down to and including Aristotle's master, Plato, philosophy perpetuated the traditional form of exposition—the cosmogonical myth, a narrative describing the birth or formation of an ordered universe. Such myths are found all over the world, in societies where science has never begun to exist. They exhibit two main patterns, singly or in combination: the evolutionary and the creational. In the one the world is born and grows like a living creature; in the other it is designed and fashioned like a work of art. The formula is familiar: "In the beginning the earth was without form and void; and darkness was upon the face of the deep." Or, in more refined language: "In the beginning was an indefinite incoherent homogeneity." The initial assumption is that the complex, differentiated world we see has somehow arisen out of a state of things which was both simple and disorderly.

The earliest Greek school at Miletus in the sixth century followed the evolutionary scheme. The original condition of things was water or mist. Cosmogony then proceeds to tell how this primitive moisture was condensed to form the solid core of earth, and rarefied into the encompassing air and the heavenly fires. Then, within this elemental order, life was born in the slime warmed by the sun's heat. This evolutionary tradition culminated in the Atomism of Democritus, towards the end of the fifth century. His system, with slight modifications, was adopted, after Aristotle's death, by Epicurus, and reproduced by Lucretius for the Roman public in the first century B.C. For Democritus the original state of things was a chaos of minute solid bodies, moving incalculably in all directions in a void, colliding, and forming vortices in which ordered worlds arise, by necessity and chance without design. There are innumerable worlds, some being formed, others falling to pieces, scattered through unlimited empty space.

The alternative pattern, preferred by Plato for moral and religious reasons, is the creational. The world is like a thing not born but made, containing evidences of intelligent and intelligible design. Necessity and chance play only a subordinate part, subdued (though not completely subdued) to the purposes of a divine Reason. For convenience Plato retained the old narrative form of exposition; but neither he nor Aristotle believed that the cosmos had any beginning in time or will ever come to an end. So Plato's myth of creation in the *Timaeus* is really a disguised analysis of the complex world into simpler factors, not a literal history of its development from a disorderly condition that once actually existed.

For Plato and Aristotle there is only one world, a spherical universe bounded by the fixed stars. Plato held that it was animated by a World-Soul, whose intelligence is responsible for those elements of rational order which we can discern in the structure. Blind necessity and chance are also at work, producing results which no good intelligence could desire; but they are in some degree subordinated to co-operate with benevolent Reason. For this Reason Aristotle substitutes a vaguely personified Nature, who always aims at some end.

Now, whichever of the two patterns be adopted—the evolutionary or the creational—cosmogony deals dogmatically with matters wholly beyond the reach of direct observation. You must, indeed, look at the world to see that there is a solid earth at the centre, surrounded by layers of water, air, and fire; but no one had observed the primitive disorderly condition, or how cosmos arose from it, and life came to be born. Nor did it occur to the ancients that their imaginative reconstruction of the past could be checked by any experimental test. For example, Anaximenes, the third philosopher of the Milesian school, held that as the primitive air or mist passed from the gaseous state to the liquid, as water, and from the liquid to the solid, as earth and stones, it became colder and also denser, more closely packed. On this showing ice ought to take up less room than water. But Anaximenes never set out a jar of water on a frosty night so as to find out how much the water would shrink when turned to ice. The result would have surprised him. It is still stranger to our minds that no critic should have thought of confirming or confuting him by this means.

This neglect of experiment is connected with the traditional form of exposition. Physical theories were stated, not as hypotheses, but as a narrative of what happened in the remotest past: "In the beginning" there was water, or mist, or qualities like hot and cold, or atoms of definite sizes and shapes. Who could decide which of these accounts was to be preferred? A physicist could do little more than accuse others of inconsistency; he could not prove his own doctrine to be true. "We are all inclined", says Aristotle, "to direct our inquiry not by the matter itself, but by the views of our opponents; and, even when interrogating oneself, one pushes the inquiry only to the point at which one can no longer find any objections" (*De Caelo*, 294 b 7). On the other hand, these early philosophers did good service by thinking out a number of alternative possibilities, some of which might bear fruit later. Atomism, which has recently borne astonishing fruit, might not have been thought of, if Democritus had not allowed his reason to outrun his senses, and assert a reality which the senses can never perceive, and no means of observation then existing could verify.

So much for differences of method. My second point is the difference in objective: what it was that the ancients were bent upon discovering.

Both types of cosmogony can be regarded as answering the question: what things really and ultimately *are*? Suppose you say that the objects we see around us are compounds of earth, water, air, and fire, and that earth, water, and fire themselves were originally formed from air, by condensation or rarefaction. You will then hold that everything now really and ultimately consists of air, in different states of density. Or you may say that everything really consists of atoms. On these lines the evolutionary type of cosmogony will declare that the real nature of things is to be found in their matter. Your philosophy will then be materialistic; and you may go on to say (as Democritus did) that the soul consists of specially mobile spherical atoms, and that all our thoughts and feelings are to be explained in terms of the motions and collisions of minute impenetrable bodies. To some this may sound fantastic; but there are still people who would like to believe something of the sort, and there are signs that the Epicurean philosophy is again becoming popular.

To this question of the real nature of things, the creational type returned a different answer. It found this real nature, not in the matter, but in the form. That was because it looked on the world as a product of craftsmanship; and the essence of such products lies in their form.

A potter is moulding clay. You ask, what is this thing

he is making? A teapot. What is a teapot? A vessel
with a spout to pour the tea through, a lid to keep it
hot, and a handle to hold the thing by without burning
your fingers. You now understand the nature of the
object in the light of the purpose dictating its essential
features. The material is not essential: you can make
a teapot of clay or of silver or of any rigid stuff that holds
liquid. The essence or real being or substance of the
thing is its form. Now suppose that the world is like a
teapot in being a work of design. Matter will then exist
for the sake of the form that is to be realized. The essence
of living creatures will be the perfect form into which
they grow. It is manifest in the full-grown tree, not in
the seed. The real nature must be sought in the end,
not in the beginning, and the end irresistibly suggests
the aim of conscious or unconscious purpose. This type
of cosmology reached its perfection in Plato and Aristotle,
in deliberate opposition to materialism.

But whether the answer be matter or form, both
types tell us what things really *are*; they do not confine
themselves to the question, how things *behave*. Here is
the second point of difference between ancient and
modern natural philosophy. At all times the quest is
for something permanent, and therefore knowable, in
the ceaseless flow of appearances. For the ancients this
permanent something is substance, whether substance
be understood as tangible material substance or as the
intangible essence of the specific form.

Aristotle takes both into account: he speaks of the
material and formal "causes" of things. Neither is a

"cause" in our sense of the word. They are the two
constituents, which answer the question: what *is* this
thing? The moderns, on the other hand, are concerned,
not so much with what things are, as with how they
behave. By a cause we mean some phenomenon or
event which regularly precedes some other phenomenon
or event, called its "effect". We are looking for those
invariable connections or sequences which are known
as "laws of nature". Such laws do not describe the
internal nature of things, but rather the constant relations between them.

Why was there this difference of objective—the
ancients defining the substance of things, the moderns
formulating sequences of events? One reason was that,
for the ancients, the pre-eminent science, setting the
pattern of all organized knowledge, was geometry.
Geometry alone had developed a method and technique
of establishing necessary truths—proving conclusions
that must be accepted by anyone who accepted the
premises. And the method of geometrical reasoning
was leading to a continual and triumphant progress in
discovery. No wonder that the search for something
certain and knowable in the physical world should follow
this brilliant example and unconsciously imitate its
methods.

Now geometry is not at all concerned to describe the
sequence of events in time. It has no use for observation
or experiment. It starts from a definition, stating, for
instance, what the triangle essentially is. It then goes
on to deduce from that definition and a few other

explicit premisses, a whole string of necessary properties of the triangle: its angles are equal to two right angles, and so on. If you can exhaust these necessary properties, you will know all that can be known about the triangle. When, with that ideal of knowledge in mind, you turn to the physical world, you will be disposed to inquire after the essential nature of visible and tangible things, and to enumerate their necessary properties, in the hope of knowing all that can be known about them. You may then develop a technique of definition by generic and specific differences. What is a man? Plainly he falls under the genus Animal, as the triangle falls under the genus Plane figure. What is the essential or specific difference, distinguishing man from other animals, as the triangle is distinguished from other figures by having three sides? Man is a biped; but so is a goose. We must add another difference, "featherless", or perhaps "rational", to distinguish man from birds. So a genus is divided into species by a method of classification, which was first elaborated by Plato and which still persists in zoology and botany. The procedure answers the ancient question: what is the essential nature of this thing?

Aristotle tries to reduce to this pattern even such questions as the cause of an eclipse of the moon. He treats "eclipse", not as an event, following upon some earlier event called its cause, but as an attribute of the moon. The moon is the subject, and when you state the fact that it is eclipsed, you are saying that it has the attribute "eclipse". If you then ask for the reason—

why has the moon this attribute?—the answer will be
the same as if you ask for a definition of eclipse. "It
is clear", he writes, "that the nature of the thing and
the reason of the fact are identical. The question "What
is an eclipse?"and its answer: 'Privation of the moon's
light by the interposition of the earth', are the same as
the question: 'Why is there an eclipse?' and the answer:
'Because of the failure of light when the earth is inter-
posed'." Thus the inquiry for the cause of an event is
reduced to inquiry for the definition of an attribute. We
ask: What is the essential nature of an eclipse? just as we
asked: What is the essential nature of a triangle or a man?

Aristotle is not setting out a sequence of two events,
one of which precedes the other and brings it about.
What moderns call the "cause"—the interposition of
the earth—is to Aristotle part of the definition of an
affection suffered occasionally by the moon.

This manner of approach has a further consequence.
When we pass from the abstract and timeless objects of
geometry to the changing things in this visible world,
we find that individual men, unlike the triangle, have
also many properties that are not essential. A man may
be tall or short, white or black, wise or foolish. Besides
the essential core of properties, without any one of which
he would not be a man at all, there is, in any particular
man, a fringe of attributes which he may or may not
have, may acquire or lose without ceasing to be human.
These attributes are called "accidental" or "contingent", as opposed to essential and necessary. If your
object is to define the universal essence common to all

men, you will rule out these accidental properties of individuals as beyond the scope of knowledge. And the words "accidental", "contingent" suggest chance—what is not determined one way or the other, but may or may not be so. The notion of chance is very obscure, and I cannot pursue it; but I believe that ancient views of the world allowed more scope for chance than is commonly recognized.

If so, that is because the ancients were not thinking of Nature, as we think, in terms of invariable laws of cause and effect. When you arrive at that notion, chance must disappear. Every event must have another event before it as cause, and before that yet another, and so on for ever. Order and Necessity will now cover the whole field, usurping the old domain of the accidental, contingent, disorderly, unknowable. So the belief in universal law led modern science to complete determinism. Miracles were not to happen. The gods were either eliminated or pushed back to an imaginary beginning, with the honorary title of First Cause—honorary, because no one really believed that there could be such a thing as a first cause. And man himself was asked to surrender the inveterate belief in his own freedom, lest he should break in upon the chain of necessary events and start a fresh and unpredictable series. In order that Nature may work like a perfect machine, man must keep in his place as a part of the machine. The ancients, in the period we are considering, were not troubled by this question of freedom, because they did not think of Nature as a perfect machine.

That word "machine" brings me to my third point: the question of the motive or driving impulse behind the two traditions of natural philosophy, the ancient and the modern. The difference in objective carries with it two different ways of looking at Nature. Scientific inquiry must select and concentrate attention upon certain aspects of the world, ignoring other aspects as irrelevant. And this selection is determined by interest, some feeling of need or desire, some value set upon this or that end in life.

Now it is a truism that the era of modern science with its mechanistic view of Nature has coincided with the era of mechanical invention, from Leonardo to Marconi. You will notice also that two of the later lectures in this series will deal with the relation of Science to Industry, in the Middle Ages and in modern times. But there is no lecture on the relation of ancient science to industry. The reason is that natural philosophy as pursued in the classical period had no bearing whatsoever on mechanical inventions. It was for this lack of interest in the means of production that the ancient philosophers were denounced in the first year of Queen Victoria by Lord Macaulay in his essay on Francis Bacon. Macaulay exalts Bacon as the apostle of modern scientific progress. A philosopher of our own day has recognized in Bacon the prophet of Big Business. There is surely some connection between the two descriptions. But listen for a moment to Macaulay's panegyric:

The chief peculiarity of Bacon's philosophy seems to us to have been this, that it aimed at things altogether different from those which his predecessors proposed to themselves. This was his own opinion.... What then was the end which Bacon proposed to himself? It was, to use his own emphatic expression, "fruit". It was the multiplying of human enjoyments and the mitigating of human sufferings....

Two words form the key of the Baconian doctrine, Utility and Progress. The ancient philosophy disdained to be useful and was content to be stationary.... It could not condescend to the humble office of ministering to the comfort of human beings.... Once indeed Posidonius, a distinguished writer of the age of Cicero and Caesar, so far forgot himself as to enumerate, among the humbler blessings which mankind owed to philosophy, the discovery of the principle of the arch, and the introduction of the use of metals.... Seneca vehemently disclaims these insulting compliments. Philosophy according to him, has nothing to do with teaching men to rear arched roofs over their heads. The true philosopher does not care whether he has an arched roof or any roof. Philosophy has nothing to do with teaching men the use of metals. She teaches us to be independent of all material substances, of all mechanical contrivances....

The ancient philosophers did not neglect natural science; but they did not cultivate it for the purpose of increasing the power and ameliorating the condition of man.... Seneca wrote largely on natural philosophy, and magnified the importance of that study. But why? Not because it tended to assuage suffering, to multiply the conveniences of life, to extend the empire of man over the material world; but solely because it tended to raise the mind above low cares, to separate it from the body, to exercise its subtilty in the solution of very obscure questions.

Finally, in a very eloquent passage, Macaulay rebukes the ancient philosophers for their failure to achieve the only practical good they aimed at: they did not "form the minds of men to a high degree of wisdom or virtue".

Whereas no one can deny that nowadays every year makes an addition to what Bacon called "fruit". "We know that guns, cutlery, spyglasses, clocks, are better in our time than they were in the time of our fathers, and were better in the time of our fathers than they were in the time of our grandfathers."

It did not occur to Macaulay that a closely similar indictment might be drawn against the religion founded by one who said that man does not live by bread alone. But more recent disciples of the Baconian philosophy have not hesitated to accuse Christianity of failing to achieve its one practical aim: to make men love one another. Both failures must be frankly admitted. But no one, as Macaulay says, can deny that modern science has produced not only better cutlery and spyglasses, but better guns, to which we can now add incendiary bombs and poison gas.

Granting all this, as indeed we must, let us consider its bearing on our question of underlying motive. If the tree is to be known by its fruits, Macaulay's description of the Baconian fruits suggests that the conscious or unconscious aim of natural philosophy since the Renaissance has been to multiply the conveniences of life and to extend the empire of man over the material world— in a word, the increase of wealth and power. Certainly this was not the aim of natural philosophy in ancient times, which, accordingly, never tried to improve cutlery, guns, and spyglasses. And the difference of aim ought to throw light on those differences of method and objective which I have outlined.

That splendid and triumphant progress towards wealth and power, which has made Europe to-day so much happier than it was in the days of Pericles or of Marcus Aurelius, has been achieved by the invention of machines, which take the work out of human hands and perform it a thousand times more quickly and efficiently. And the construction of machine tools means enlisting the tremendous forces provided by Nature. The first forces to be utilized are the passive forces of weight and pressure exerted in the natural motions of air and water—the wind that fills the ship's sails, the stream that drives the water-wheel. It is significant that the pioneers of science in the sixteenth and seventeenth centuries, notably Galileo and Newton, were specially interested in the laws of motion and gravity, which they were the first to formulate. Later came the much more powerful active energy released by combustion. After taming earth and air and water, man harnessed fire to his engines of production.

You cannot effectively enlist these natural forces until you know a good deal of their working apart from human control. So a science ultimately bent on the fruits of power and wealth will find it useful to regard Nature itself as a machine of unsuspected complexity. The first task will be to take this machine to pieces, and to grasp the relation of one part to another, and how each part behaves. A machine must work with the greatest possible order and regularity. So the mechanical philosophy of Nature looks always for those invariable sequences of cause and effect which are the secret of its uniform behaviour. The results discovered are then transferred to machines contrived by human ingenuity. These machines will not work unless the behaviour of the forces employed has been apprehended correctly. Knowledge that is turned to practical use is constantly checked by results. Hence a whole technique of exact observation and cautious experiment will be invoked to make sure that the Jinn conjured out of the jar will serve the magician's will and not tear him to pieces.

Let us now contrast with this attitude towards Nature as a source of mechanical power the attitude of the ancient Atomists as formulated by Epicurus and reproduced by Lucretius. I take Epicurus for two reasons. First, Atomism has a closer kinship than any other ancient system with modern physics. Secondly, Macaulay saw in Epicureanism the one sect which ought not to have merited Bacon's condemnation. "The Epicurean who referred all happiness to bodily pleasure and all evil to bodily pain might have been expected to exert himself for the purpose of bettering his own physical condition and that of his neighbours. But the thought never seems to have occurred to any member of that school." True; but why was the Epicurean so indifferent?

It was, perhaps, not the fault of the philosophers that the ancients had so few machines driven by non-human force. The simple fact is that power for anything that could be called large-scale industry was then supplied by slaves—"living tools' , as Aristotle called them. It is estimated that when Pericles died the population of

Attica was divided into a little more than 200,000 freemen and 115,000 slaves. Abundant slave-labour makes it unnecessary to enlist non-human force in industry. There was thus no economic pressure driving men to study Nature as a source of mechanical power. The philosopher or man of science of any age, if he cares for knowledge and not for riches, has no personal motive to invent machinery for production. In a society where industry already has all the power it needs, you will not find the man of science installed in a factory, and devoting years of research to devising a process of making cotton fabrics that will not crease when folded.

But Epicurus was not merely free from any external pressure to discover the energy latent in his atoms. If he had held the key to unlock that energy and harness it to ever more powerful machines, then, like Prospero, he would have abjured that magic, and, deeper than did ever plummet sound, have drowned his book. He believed that human happiness depends, not on intense and varied pleasures, but on untroubled peace of mind; and that the pursuit of wealth and power had made man less happy even than his primitive ancestors before they found out the use of fire and the working of metals. As Lucretius says;

If a man would order his life by a true principle, for him a frugal subsistence joined to a contented mind will be great riches; for he whose needs are small will never be in want. But men desired to be famous and powerful, hoping that their fortunes might rest on a firm foundation, and wealth might enable them to lead a tranquil life. But all in vain. . . .

Man labours to no purpose and wastes his life in fruitless

cares, because he has not learnt what is the true end of possession, and up to what point true pleasure goes on increasing. This by slow degrees has carried life out into the deep sea, and stirred up from their lowest depths the mighty billows of war.

If my train of thought has been sound, this difference of motive and consequent interest—in the last resort a question of human values—lies at the root of the other differences we have noted. The arts of peace, as they were called in Macaulay's day, are now openly described in terms of the art of war. In Russia a party of labourers, who have painfully learnt what hard work means, are said to be despatched as "shock troops" to "the agricultural front". All wars, as Plato remarked, are made for the purpose of getting money and the material things that money can buy. It is now admitted that industry at home and commerce abroad are a warfare waged for the same purpose. There is also the class war, to decide whether the money and the goods shall go to the rich or to the poor. I have suggested that, for economic reasons, the ancient study of Nature was not drawn into this perennial struggle. So it was suffered to remain as part of the pursuit of peaceful wisdom and of a happiness independent of wealth and even of material comfort. The fruits it gathered from the Tree of Knowledge were not the Baconian fruits of utility and progress.

II. FROM ARISTOTLE TO GALILEO

by

SIR WILLIAM CECIL DAMPIER

Fellow of Trinity College, formerly Secretary of the
Agricultural Research Council

FROM ARISTOTLE TO GALILEO

To understand the predominance in later mediaeval times of the Athenian school of Greek philosophy, we must trace earlier modes of thought, which, in some ways, were nearer to our own. Beginning with the geometry of Thales of Miletus (c. 580 b.c.) the speculations of the early Ionian philosophers culminated 150 years later in the atomic theory of Leucippus and Democritus, preserved for us in the philosophy of Epicurus and in the poem of the Roman Lucretius. Primarily the theory was a bold attempt to explain the ultimate nature of different kinds of matter in terms of eternal atoms, many in size and shape but identical in substance, moving for ever unless opposed, and colliding with each other to form substances and worlds. The atomic theory also formulated the principle of cause and effect—"everything happens with a cause and by necessity". The Atomists saw too that to admit the qualities of bodies as fundamental entities would stop all further analysis. Therefore, Democritus said, "according to convention there is a sweet and a bitter, a hot and a cold, and according to convention there is colour. In truth there are atoms and a void". Unaided sense-perceptions do not reveal to us reality.

The atomic theory of the Greeks was, as I have said, nearer to modern views than the ideas that followed it, but, since it was not based firmly on observation and experiment, it failed to survive. Unlike the men of Ionia, the Athenian philosophers were more interested in the mind of man than in external nature. Plato's cosmos was a living organism with soul and body. He prized mathematics as a gentlemanly, deductive science, but roundly condemned experiment as either impious or a base mechanic art. Natural objects as individuals, he held, are changeable and uncertain, but can be grouped in classes. The classes may be represented by the mind as ideal, primary types, and it is to these types alone that definitions and reasoning can be applied. Hence Plato concluded that these types or forms are the only realities. Such an outlook, while favouring mathematics, and having affinity with modern scientific philosophy, was unfavourable to the development of primitive experimental science.

Aristotle was a pupil of Plato, and accepted in a modified form his master's idealistic "realism". Nevertheless, he allowed reality also to individuals, and thus found a philosophic base for the beginnings of natural science. He used it well, especially in the observational science of zoology, where no advance on his results was made till modern times. Aristotle was the greatest biologist, the greatest logician and the greatest systematizer of knowledge, of the ancient world.

In physics he was less successful. He argued that in a vacuum all bodies heavy or light must fall equally fast. This equality he regarded as inconceivable, and therefore he concluded that no vacuum can exist. With the idea of empty space, he rejected all the other allied concepts of the atomic theory, arguing that, if all substances were composed of the same ultimate material, they would all be heavy by nature, so that nothing would be light in itself or rise spontaneously. His failure, of course, is due to his belief that the inconceivable is fundamentally impossible, and to his ignorance of Archimedes' later discovery of density or specific gravity. He rejected also the Atomists' belief that bodies once set in motion continued to move till opposed. Hence he taught that to keep the heavens revolving an Unmoved Mover must continuously be at work. He held that celestial bodies were divine, incorruptible, unlike those on earth. Nevertheless, he regarded the earth as the centre of the Universe, a view later incorporated in the astronomy of Hipparchus, and handed on to the Middle Ages in the writings of Ptolemy.

Rejecting the atomic theory, Aristotle returned to the older view that the essence of matter was to be found in its qualities. He imagined four primary qualities hotness and coldness, wetness and dryness, which united in binary combination to form the four elements earth, water, air and fire. These, in varying proportions, built up the different kinds of matter. Thus, to Plato and Aristotle, qualities familiar to the human mind were the essence of matter, and not the atoms of Democritus, unconcerned with man, his understanding or his welfare. With Democritus' atoms, they were able to banish also his unwelcome materialism and determinism.

Aristotle was also the creator of formal logic, a tremendous achievement. He applied his results to the theory of science, successfully using mathematics to supply examples. But logical syllogisms are useless for experimental science, which needs inductive discovery, and not deductive proof from accepted premisses. It is probable that Aristotle's deductive theory of science did much to retard the growth of natural knowledge.

In contrast with the humanist, metaphysical philosophers of Athens, were the later physicists like Aristarchus of Samos, an ancient forerunner of Copernicus, and Archimedes of Syracuse, who was primarily a geometer, but also the founder of mechanics and hydrostatics. These men, instead of building general philosophic systems and deducing particulars therefrom, attacked definite and limited scientific problems with success. The same more modern methods are also seen in many of the inquirers of Alexandria, which succeeded Athens as the centre of the intellectual world.

But the early Fathers of the Church were attracted by the more mystical ideas of Platonism and neo-Platonism, and put together the first Christian synthesis of these ideas as interpreted by St Augustine, mingled with elements of contemporary mystery religions and traces of earlier heathen cults. This was the view of the world, which, surviving through the Dark Ages, remained predominant in early medieval times. Compared with it, Aristotle would have been scientific, but the full texts of Aristotle were lost; his works were known only in abstracts and compendiums, and during these centuries had little effect. In this period the elements of Greek science were best preserved in the Arabic schools of Persia, Iraq and Spain, where some new discoveries were also made.

In the revival of learning which marked the thirteenth century, the works of Aristotle were rediscovered, and full translations into Latin were made, first from Arabic sources and then direct from Greek manuscripts. A far wider outlook than anything then available both in philosophy and science was thus opened out, and it needed a bold effort to reconcile Aristotle's ideas with contemporary Christian dogma. At first those ideas suffered ecclesiastical censure, but they proved irresistible and in 1225 the University of Paris agreed that Aristotle's works should be studied.

The new synthesis was made chiefly by Albertus Magnus of Cologne and his pupil St Thomas Aquinas. The great scheme of Scholasticism which they constructed was meant to comprise the whole of knowledge, both sacred and profane. The accepted theology based on Scripture and interpreted by the Church was amalgamated with secular learning derived mainly from Aristotle. Accepting the premises, great ingenuity and close reasoning appeared in the methods by which the Schoolmen reached their deductions, and, in so far as they upheld the intelligibility of the Universe and the supremacy of reason, they prepared the way for modern science.

On the other hand they incorporated in their scheme a good many of the errors from Aristotle's physics; they accepted the Ptolemaic system of astronomy, although St Thomas himself regarded it merely as a working hypothesis. Thus the geocentric theory became part of the Thomist philosophy; as man was the object of creation, so the earth was the centre of the Universe. Again, the Schoolmen adopted Aristotle's beliefs that to maintain motion the constant exertion of force is necessary and that the planets and stars were driven in their courses by an Unmoved Mover.

To Thomas Aquinas, creation is a procession of all creatures from God; ideal types are forms pre-existing in the Divine Mind, and the meaning of individuals is to be sought in determinate matter, a new solution of Plato's old problem of universals.

The whole scheme was permeated with Aristotle's preference for the logical, deductive type of knowledge, all the more so because the Schoolmen felt they had founded their work on the impregnable rocks of Holy Scripture, the Fathers of the Church and the rediscovered books of Aristotle. On such authority the premises had to be accepted, and the deductions were made by the acutest minds of the thirteenth century. Thus arose a rational, self-contained synthesis of all existing knowledge, so closely knit together that an attack on any one scientific deduction became an attack on the whole, including the Christian verities. The few medieval experimentalists, Roger Bacon chief among them, pitting isolated facts against such a complete rational philosophy, were quite ineffective. Though later on more damaging attacks were made from the philosophic side, speaking broadly Thomist Scholasticism dominated the world for more than three hundred years. It was a theological but understandable scheme, interpreted in terms of man's sensations, desires and welfare, with heaven shining just above the sky and hell rumbling not far beneath the ground. On its tremendous authority, men accepted a congenial anthropocentric philosophy, and, for the most part, were content.

In the early fifteenth century, a growing taste for classical literature began to attract Greeks from the East, who, from their knowledge of the modern tongue, could teach its ancient prototype. This migration was hastened by the capture of Constantinople by the Turks in 1453. Thus the language of ancient philosophy and science became familiar to Western scholars after a lapse of eight hundred years.

At first the revival was chiefly literary, but, in contrast with authoritative Scholasticism, the spirit of humanism was one of free inquiry. First in the cities of northern Italy, and then across the Alps, this freer spirit spread, the beginning of the Renaissance.

It is often said that the sudden change of outlook associated with that time is so great as to be inexplicable. But, as I have pointed out elsewhere, although a number of simultaneous movements while isolated produce a total effect the sum of their individual effects, when they have spread enough to overlap, they may act and react upon each other, and their total result become cumulative. So it was with the Renaissance. The recovery and gradual assimilation of Greek learning, the invention of movable type which made printing possible, the voyages of discovery, which opened up Africa and revealed America, were some of the factors involved. Especially the gold and silver from Mexico and Peru produced an economic revolution in Europe. The value of money fell, that is, prices rose. Customary rents and other fixed payments became less burdensome, trade and industry made profits, flourished and expanded, and rapidly increasing wealth gave greater leisure and opportunity for study.

The first task was to recover the learning of the Greeks—far beyond that of intervening ages—to get behind the distorted version given by Thomist Scholasticism to Plato and Aristotle, the originals. There was danger that another authoritative philosophic orthodoxy might arise. But we now know that, by the end of the fifteenth century, a new idea had dawned in intellectual circles in Italy, the idea that knowledge unknown to the Greeks might be discovered by observation, reasoning, and experiment. Mathematics, astronomy, anatomy, were being studied by men whose shadowy figures have come down to us in the writings of others.

The most interesting records of the time are the manuscript note-books of that universal genius Leonardo da Vinci, which have been copied and printed in our own day. Leonardo studied the works of the Greeks and especially those of Archimedes, the most modern of the ancients, but, as a painter, sculptor, architect, and engineer, he also approached science from the practical side, and realized that it was more important to find out what really happened than to read what Aristotle thought about it. He studied the laws of optics and the structure and use of the eye, other details of human anatomy, and the flight of birds. In formulating the principles of statics and dynamics, he proved the law of the lever as the primary machine, and foreshadowed the idea of inertia. He used the right experimental method, and understood the right experimental philosophy. Mathematics give certainty within their own realm; they are concerned only with ideal mental concepts. But practical sciences are vain and full of error unless they arise from experiment, though it is well if mathematical reasoning can then be applied and a final, clear experiment made. If Leonardo had written and published books on his work, modern science would have begun long before Galileo.

Next let us turn to Copernicus (1473–1543). The object of his work was to find some simpler scheme of cosmogony than that given by the cycles and epicycles of Hipparchus and Ptolemy. The philosophic idea which underlay this object is of interest. Behind the predominant Aristotelianism of Thomas Aquinas there had survived traces of Augustinian Platonism, containing pre-Platonic elements which conceived the Universe in terms of a mystical harmony of numbers or of geometrical arrangements of units of space. Thus these later Platonists were always seeking mathematical relations in Nature; the simpler the relation the nearer it was to reality. Moreover, some of the ancient philosophers including the Pythagoreans believed the earth to move round a central fire, and it needed no great change in ideas to see that the central fire was the sun.

In the early sixteenth century there was a revival of this mathematical Neo-Platonism in Italy, and in Italy Copernicus spent six years. He says: "according to Cicero, Hicetas thought that the Earth moved... according to Plutarch others held the same opinion.... I myself also began to meditate on its mobility." Copernicus then accepts the sun as the centre of all things and the earth as one of the planets. He describes in detail this solar system and ends: "We find therefore under this orderly arrangement a wonderful symmetry in the Universe, and a definite relation of harmony in the motion and magnitude of the orbs."

The Copernican theory won its way very slowly, for it involved a revolution in thought. The earth sank from its proud position as the centre of the Universe to a humbler place among the planets. Ecclesiastics scented danger, and what would now be considered the best scientific opinion of the day was definitely against the new theory. If the earth spun on an axis, would not loose stones fly off, perhaps the earth itself disintegrate? If it moved round the sun, would not the stars appear to shift, unless indeed they were so far off that the distances became ridiculous if not inconceivable? Besides, was not the authority of Aristotle against Copernicus? Though a few mathematicians were convinced, the majority doubted.

Pythagorean and Platonic influences too underlay the work of Kepler. His three "laws" were only the survivors of the many attempts he made to find harmony in planetary motions. He accepted the Copernican system with enthusiasm as giving greater mathematical simplicity and harmony than any other view, being convinced that God created the world in accordance with the principle of perfect numbers, so that the harmony or music of the spheres is the true cause of their movements.

There the subject remained till Galileo Galilei (1564–1642) heard in 1609 that a Dutchman had invented a telescope. Galileo at once constructed a similar, and then a better, instrument which magnified 30 diameters, and turned it to the sky. At once discovery followed discovery. The surface of the moon was not, as Aristotle and the Schoolmen held, smooth and perfect, but rugged with broken mountains and valleys; countless new stars came into view; and the planet Jupiter, with its attendant satellites, showed a smaller scale model of Copernicus' solar system. The professor of philosophy at Padua refused to look through Galileo's telescope, and his colleague at Pisa lectured before the Grand Duke against the new discoveries. But Galileo had confirmed with sensible facts, which anyone could verify, Copernicus' views, hitherto based only on philosophic grounds of mathematical simplicity.

Galileo's second and greatest achievement was his work in dynamics. Men's ideas on motion were a confused medley of unco-ordinated observation and Aristotelian theories. Bodies were thought to be intrinsically heavy or light, and to rise or fall with velocities proportional to their heaviness or lightness, because they sought their natural places with varying power. Repeating a forgotten experiment of Stevinus, Galileo dropped a ten pound weight and a one pound weight from the top of the Leaning Tower of Pisa. The horrified onlookers had to confess that, heavy and light, they reached the ground together.

Galileo felt that the movement of bodies on the earth, like that of the planets, should be explicable in mathematical terms, and he set himself to discover the relations that held—to find out not *why*, but *how* things moved. A falling body moves with increasing speed. What is the law of the increase? His first guess was that the velocity might vary as the distance fallen through. This, he found, involved an inconsistency, and he then tried the hypothesis that the velocity increased with the time of fall; he deduced its consequences, and compared them with the results of experiment.

A body falling freely soon moves too fast to be measured with the simple implements available to Galileo, and he turned to the use of inclined planes. He found that a body falling down an inclined plane acquired the same velocity as though it had fallen through the same vertical height. His measurements agreed with his hypothesis that the velocity varied with the time, and its mathematical consequence that the space described increases with the square of the time.

Again Galileo found that, after running down one plane, a ball will run up another to a height equal to that of its starting-point whatever be the slope, provided friction be negligible: it is the height alone that matters.

Thus, if the second plane be horizontal, the ball will run along it with uniform velocity till checked by friction or some other opposing force. Here we have the refutation of Aristotle's belief that the continued exertion of force is necessary to maintain motion, and an Unmoved Mover to drive the stars: even Kepler thought the planets were somehow kept moving by the sun working through an aether. Galileo's discovery proved that it was not motion, but a change in velocity or direction that needed the action of a force. Till Galileo's discovery it was impossible even to formulate aright the problem of the solar system, but now Galileo had cleared the way for Newton.

An important new concept lies implicitly in Galileo's experiments—the concept of inertia or mass. Leonardo had said "every body has a weight in the direction of its movement", and Galileo's ball, rolling on a horizontal plane, could only be set in motion or stopped by force, and the heavier the ball the greater the force needed. Though Galileo did not explicitly point it out, there is here something connected with but distinct from weight, first definitely separated therefrom by Balliani, a Captain of Archers at Genoa, who distinguished *moles* from *pondus*.

Newton defined mass as the product of density and bulk, and force as measured by the change of motion, i.e. rate of change of momentum. Now, as Mach pointed out in 1883, this lands us in a logical circle, for density can only mean mass per unit volume. We can escape from the circle by defining mass as inversely proportional to the accelerations produced in two bodies either acting on each other, say when connected by a coiled spring, or by forces equal as measured by muscular effort. In either case, the justification of the concept of mass as a constant quantity is obtained by its use in countless dynamical problems and the concordance of the deductions with observation. Mass proved to be constant till the days of J. J. Thomson, G. F. C. Searle and Einstein.

The relation between mass and weight is also implicitly contained in Galileo's experiments. Acceleration is measured by force divided by mass. With falling bodies, the forces acting are the weights. Since heavy and light fall together, the accelerations are equal, and the weights must be proportional to the masses. This result was proved explicitly by Newton by experiments with pendulums. It could not be reached by any juggling with formulae. It was an experimental result, and indeed a very surprising one.

Galileo, as said above, showed that the velocity of a falling body increases with the time, that is, $v = \alpha t$ or $mv = ft$, the law used by Newton. But from $v = \alpha t$ we get $v^2 = 2\alpha s$, where s denotes the space traversed, and had Galileo first happened on the relation that the square of the velocity varied with the distance, he would have regarded the relation $v^2 = 2\alpha s$ as fundamental, and reached the conclusion that $fs = \frac{1}{2}mv^2$, the kinetic energy relation assumed by Huygens. Thus, as Mach pointed out, only one fundamental principle underlies the dynamics of Galileo, Huygens and Newton. The different modes of presentation were historical accidents.

It will be seen how far Galileo departed from the ideas of motion held by Aristotle and the Schoolmen, who were wont to derive them from metaphysical assumptions about ultimate causes. They analysed motion in terms of "substance" with the help of vague notions such as "action", "efficient cause", "end" and "natural place". As Euclid and his predecessors converted geometry into mathematics, and Hipparchus, Copernicus and Kepler reduced astronomy to geometry, so Galileo did the same for terrestrial dynamics. First giving definite mathematical form to the old concepts of space and time, somewhat unimportant categories in Aristotelian thought, Galileo gave them that primary and fundamental character which they have held since his day. These clear underlying concepts enabled Galileo to formulate and then solve his dynamical problem.

In our own times idealistic philosophers have blamed the change of outlook produced by Galileo as the cause of some of our metaphysical difficulties. As long as Aristotelian ideas held, the problem of knowledge did not arise. The physical world was described—confusedly and inaccurately it is true but still described—in humanist ideas, so that explanations, such as they were, appeared in terms natural to human minds. But Galileo's fundamentals were mathematical time and space, and matter moved by forces in accordance with mathematical relations. He was thus led back to the theory of the Greek Atomists, who explained the world in terms of particles moving under immutable necessity, caring nothing for man, his understanding, his desires or his welfare. But, if this, or anything like it, be true, how can the non-material mind of man grasp material happenings? There seems no point of contact. Galileo was content rather than make guesses "to pronounce that wise, ingenious and modest sentence, 'I know it not'". The problem indubitably exists; it was hidden and obscured by Scholasticism; and, first of the moderns, Galileo brought it clearly into view.

Again, Galileo recognized that what have been called secondary qualities—colour, taste, smell, etc.—are merely sensations in the mind of the observer, and not essential properties of the substance concerned, as he thought were the primary qualities, such as extension, movement or rest, and existence in definite space and time. In the eighteenth century Bishop Berkeley went further, and argued that the so-called primary qualities also are relative to the mind of the observer. We thus arrive at the extreme form of idealism which holds that nothing really exists unless it is observed by a sentient being. If after nothing we insert "as known to our senses", this view of reality is hard to refute. The philosophy is that described in the famous limerick, which I should not venture to quote had it not been written by an eminent Oxford theologian:

> There was a young man who said, "God,
>   To you it must seem very odd
> That a tree as a tree simply ceases to be
>   When there's no one about in the Quad."

Some of you may not have heard the less well-known,
but unanswerable, rejoinder:

> Young man, your astonishment's odd,
> I AM always about in the Quad,
> And that's why the tree continues to be
> As observed by, Yours faithfully, God.

Thus Theism clearly gives us one solution of the problem.

Galileo's great discovery that terrestrial movement
can be described in mathematical terms opened the
tremendous advances in science of the Newtonian epoch.
On the hypothesis that particles attract each other in
proportion to the product of their masses and inversely
as the square of their distance, Newton proved that a
sphere attracts as though all its mass were concentrated
at its centre, and was then able to link the fall of stones
or apples to the ground with the majestic sweep of the
moon in her orbit, and extended Huygens' rationaliza-
tion of circular motion by the demonstration that an
inverse square law of force would explain the elliptic
orbits of the planets and Kepler's laws which describe
their motion. This mighty synthesis of Galileo, Huygens,
and Newton converted the confused heavens of Aristotle
into a calculable dynamic machine.

Newton himself interpreted his cosmogony in a
religious sense. "This most beautiful System of the
Sun, Planets and Comets", he wrote, "could only
proceed from the Counsel and Dominion of an intelligent
and powerful Being...." God, he says, "endures for
ever and is everywhere present, and by existing always
and everywhere, He constitutes duration and space".

While this attitude of mind was carried further by some of Newton's disciples like Bentley and Samuel Clark, a very different interpretation was given to Newton's work by Voltaire and the French Encyclopaedists. They regarded it as giving a complete mechanical explanation of the Universe, and, transferring the results from science to metaphysics, concluded that ultimate reality was mechanical too. Thus the French eighteenth-century Encyclopaedists led to the French and German materialists of the nineteenth century. All this can be traced back through Newton to Galileo, and, as I have said, idealistic philosophers blame them for it. Professor E. A. Burt writes:

Newton's authority was squarely behind that view of the cosmos which saw in man a puny irrelevant spectator (so far as a being wholly imprisoned in a dark room can be called such) of the vast mathematical system whose regular motions according to mechanical principles constituted the world of nature. The gloriously romantic universe of Dante and Milton, that set no bounds to the imagination of man as it played over space and time, had now been swept away. Space was identified with the realm of geometry, time with the continuity of number. The world that people had thought themselves living in—a world rich with colour and sound, redolent with fragrance, filled with gladness, love and beauty, speaking everywhere of purposive harmony and creative ideals—was crowded now into minute corners in the brains of scattered organic beings. The really important world outside was a world hard, cold, colourless, silent and dead; a world of quantity, a world of mathematically computable motions in mechanical regularity. The world of qualities as immediately perceived by man became just a curious and quite minor effect of that infinite machine beyond. In

Newton the Cartesian metaphysic, ambiguously interpreted and stripped of its distinctive claim for serious philosophical consideration, finally overthrew Aristotelianism and became the predominant world-view of modern times.

But Newton, as we have seen, would have repudiated any such interpretation, and we shall presently find another way of avoiding the difficulty.

If Galileo, seconded by Newton, banished Aristotle from dynamics and astronomy, his influence still survived in other realms of science and philosophy. In chemistry, his four elements disputed the ground with a more recent hypothesis which regarded the "essences" or "principles" of salt, sulphur and mercury as the base of things. Robert Boyle in *The Sceptical Chymist* argued in favour of unalterable atoms, which survive different chemical combinations, rather than Aristotelian elements or Spagirist principles. Here we have an early form of the theories of the persistence of matter, demonstrated by Lavoisier, and of the existence of chemical atoms, set forth in detail by Dalton; it is the extension of Galileo's ideas from dynamics to chemistry. The connected theorem of Huygens similarly was extended by Joule as the conservation of energy to cover forms of energy other than dynamic. By the middle of the nineteenth century, the ideas started by Galileo reigned supreme in physical science. In biology the corresponding mechanistic views were never so universally accepted; there have always been alternating periods of vitalism, in some form, and mechanism, though it was the *Darwinismus*, developed in Germany, that proved an effective ally to physical atomism in producing a wave of materialism.

Thus the nineteenth century saw the peak of the philosophy based on the physics which began with Galileo. The men of science, on whose work it was founded, were not interested in the contemporary metaphysical idealism due to Kant and Hegel, which, in its turn, ignored contemporary science. The experimentalists held, practically universally, a common-sense realism, which assumed that the metaphysical reality underlying phenomena was revealed by the scientific relations discovered in the laboratory. They only began to doubt when Mach in 1883, speaking in language they could understand, revived the ideas of Locke, Berkeley, and Hume, and pointed out that science does but construct a model of what our senses tell us about Nature, and that mechanics, far from being necessarily the ultimate truth about reality, is but one aspect from which that model can be regarded—one section of the model cut in a plane to suit our convenience. Other aspects—other sections—chemical, physiological, psychological—are equally important, equally fundamental.

This phenomenalism is, I think, the best way to face the difficulties which some men have professed to see in the science founded by Galileo. As a map or a chart represents the surface of the land in a systematic but conventional manner, so science represents the reality which underlies it. The map or chart does not show the land as we see it, like a picture or photograph, but it is consistent within itself, and it can be used safely and confidently as a guide by travellers. So the model which science constructs is self-consistent; it can be used as a guide to practical life, and as a means of predicting future physical events with a high degree of probability. But science, dealing only with these models and forms in which we group the relations between phenomena, does not within its own realm reveal or even touch reality. We are perhaps returning towards Plato's old theory of ideas. When we discover scientific relations—Laws of Nature, shall we say—they connect together ideal forms, and it is between such concepts that the relations hold. Plato argued, as I said at the beginning, that such ideal forms were the only reality, and thus arrived at idealistic realism. To us the models are the subject-matter of science—scientific, but not metaphysical or fundamental reality. Between them scientific determinism holds good. The error comes in when that determinism is transferred to the unknown concrete reality from which those concepts have been extracted.

The whole mystery that mankind has to face undoubtedly involves the problem of the nature of reality, but that is a metaphysical not a scientific problem. The fact that a consistent model of Nature can be put together by science is a valid metaphysical argument in favour of the view that some reality, corresponding in some close way to the model science makes, lies hidden beneath phenomena, but science does not directly reveal that reality as we used to think.

In the laboratory, as in practical life, there is no
room or time for philosophic doubt, but in periods of
reflection it is well to remember the purely conceptual
nature of science when based solely on its own induc-
tions.

III. FORTY YEARS OF PHYSICS

by

THE LATE LORD RUTHERFORD

(revised and prepared for press by

J. A. RATCLIFFE)

i

THE HISTORY OF RADIOACTIVITY

NOTE

THESE two lectures were delivered by the late Lord Rutherford. He had intended to write them in a form suitable for this book, and for that purpose it had been arranged that a stenographer should be present, and a verbatim account of the lectures prepared.

I have been asked by the Editors of this book to prepare these notes for publication. Those who knew Lord Rutherford's lectures will remember how far he relied on the force of his personality at the lecture table to convey his meaning, and they will realize that a mere transcript of the shorthand notes would not be in any way suitable for this book. I have therefore rewritten the lectures in a connected form, keeping as closely as possible to Lord Rutherford's own words. In some places the wording is entirely new, in others it follows very closely that of the original notes.

It is hardly necessary to add that I, alone, must accept entire responsibility for any errors in matters of fact, and for any too wide divergence from the original notes.

J. A. R.

The Cavendish Laboratory
Cambridge

THE HISTORY OF RADIOACTIVITY

I PROPOSE to give two lectures, the first on the development of ideas in the subject of radioactivity and the second on the present ideas of the structure of atoms. I think that the Committee organizing these lectures has been very wise in starting off with the history of science in our age and in drawing the dividing line in the year 1895, because that year marks a clear-cut division between what we call the old, or classical, and the new, or modern, physics. It was in that year that Röntgen made the far-reaching discovery of X-rays, a discovery which had in itself and by its consequence an enormous reaction on the advance of science. I myself was fortunate in that I came to the Cavendish Laboratory to work with Sir J. J. Thomson in that transition year 1895, and I should first like to tell you something of the attitude of scientific men at that time.

Let us briefly consider what we physicists were sure of at that date. First of all, there was the famous electromagnetic theory of Maxwell which had related light and electrical vibrations, so that light was believed to be nothing more than a form of electric wave transmitted through space. From this it followed that atomic spectra, such as the bright line spectrum emitted by hydrogen when subjected to an electric discharge, were forms of electric vibration, and therefore presumably produced by a vibration of some electric charge. For this reason many theoreticians, such as Sir Joseph Larmor and Lorentz, took the view that the atom must contain electric vibrators, although they had no idea at first whether these were positively or negatively charged.

Another generally accepted theory was the kinetic theory of gases, which supposed that the properties of gases could be explained by the motion of molecules, and, as you know, it was possible, from certain experimental results, to deduce the number of molecules in a cubic centimetre of a gas, and to estimate the size and the weight of the atoms. At this period, however, the numerical estimates made by various experts from time to time were very varied and we could only rely very roughly on the data concerning the mass or size of the atom. The reason for this uncertainty was partly that the calculations of kinetic theory were very rough and incomplete, and partly that the experimental data were not very reliable.

Most of you will not be surprised to hear that we believed in the kinetic theory and the molecular constitution of matter, but there is one point that the young student of to-day is liable to forget, and that is that the atomic nature of electricity was also generally accepted at this time. It is true there were no clear-cut experiments leading to the idea, but it was accepted as a result of the famous deductions made by Faraday many years before, from experiments on electrolysis. Most of the credit of bringing it before the public should go to

Dr Johnstone Stoney of Dublin, a philosopher whom I knew personally. He it was who saw that there must be a fundamental unit of charge carried by the hydrogen atom in the electrolysis of water, and in giving a name to that charge he coined the word electron, now applied universally to that charge.

We must now consider the state of knowledge in those branches of chemistry with which we shall be concerned to-day. As a result of centuries of industrious work the chemists had succeeded in separating and refining the great majority of the elements, and the idea had arisen that the atoms of a particular kind of matter were all made on the same pattern. They were unchangeable and indestructible, and they would last for ever, or as long as any chemical knowledge would last. Although the old idea of the solid "billiard-ball" atom had been completely discarded by the end of the last century, the chemist still felt confident that with the methods at his disposal the atoms were unchangeable and definitely indestructible. Occasionally someone thought he had transformed one kind of atom into another, but it had always been possible to prove him wrong.

There had been developing at the same time that great generalization known as the periodic law, by which the properties of the elements were related to their positions in a list of atomic weights. The more philosophical of chemical men instinctively felt that this involved the view that atoms were either similar structures or in some way all made up from some more elementary material. But the ideas were vague, and the true

meaning of the periodic law was not understood until
another ten or fifteen years had passed.

Now I come to the beginning of my story. Few of
you can possibly realize the enormous sensation produced by the discovery of X-rays by Röntgen in 1895.
It interested not only the scientific man, but also the
man in the street, who was excited by the idea of seeing
his own inside and his bones. Every laboratory in the
world took out its old Crookes' tubes to produce X-
rays, and the Cavendish Laboratory was no exception.
These old tubes of Crookes showed that cathode rays
have the power of causing brilliant phosphorescence in
a great number of substances, and it was also observed
that X-rays appeared to come from the points which
were struck by the rays. This led many people to think
that X-rays might be connected with phosphorescence
in some way, perhaps that phosphorescent substances
might emit X-rays. A number of observers on the
continent did experiments on this subject, among others
Henri Becquerel of Paris. His father, a professor before
him, had been very interested in phosphorescence,
particularly in measuring its duration, and he had also
been interested in the rather unusual properties shown
by uranium compounds. Henri helped in his father's
work and fifteen years before, in 1880, he had amused
himself by making some crystals of the double sulphate
of uranium and potassium, which glowed beautifully
when exposed to light. In his search for a connection
between phosphorescence and X-rays Becquerel placed
a number of phosphorescent substances, enveloped in

black paper, over a photographic plate, but his results were entirely negative. It then occurred to him to try his crystals of uranium salt. He first exposed them to light, so as to make them phosphoresce, and then wrapped them in black paper and placed them over a photographic plate. After an exposure of several hours and development, a distinct photographic effect was observed. The experiment was repeated with a thin piece of glass between the uranium salt and the photographic plate in order to cut off effects due to possible vapours, but the photographic effect was again obtained. At first Becquerel assumed that the emission of rays which could penetrate the black paper was in some way connected with the phosphorescence, but later he showed that the effects were just as marked if the uranium salt had previously been kept in the dark for several weeks so that there was no sign of phosphorescence. He later showed that all the salts of uranium, and even the metal itself, have the power of producing radiation which penetrates black paper. In this way he discovered the phenomenon which to-day we call Radioactivity.

We now come to a name with which you are all familiar, that of Mme Curie. She started to investigate the activity of various substances by examining the rate at which the radiations would discharge electrified bodies placed in their neighbourhood. She found that pitch-blende, and some other minerals, produced an effect greater than that of pure uranium, and she concluded that these minerals must contain some substance which was even more active than uranium. She therefore analysed the mineral chemically, going through the ordinary processes of chemical separation and at each stage retaining that portion which showed the greater radioactivity. She found two very active substances: one which was chemically similar to bismuth she called Polonium, and the other, similar chemically to barium, she named Radium.

The amount of radium in any of the radioactive minerals is very small, of the order of 1 part in 10,000,000, but by working with tons of the original mineral Mme Curie was able to prepare enough pure radium bromide for her to determine the atomic weight of radium and to show that it had a definite spectrum, in other words to show that it behaved chemically like an ordinary element.

We are indebted to Dr Giesel, chemist of the Chinin-fabrik, Braunschweig, for first putting preparations of nearly pure radium salt on the market. It is said, I do not know with what truth, that he had succeeded in separating radium a little earlier than Mme Curie, but, since he had used her methods and his work was a direct consequence of hers, he had, with proper scientific generosity, refused to claim any credit for this. However that may be, the work had an important consequence, for his interest in these substances led him to put pure radium bromide on the market at £1 a milligram. I bought 30 milligrams and Ramsay did the same. A little later it cost £12 a milligram.

The discovery of radium was of the greatest importance to science, chiefly because its activity was so great, more than a million times that of uranium, that it could not be explained away as a small secondary effect. The fact that it had a long life (1600 years) and that it was easily separated chemically, also added to its importance.

It is interesting to look back and think what would have happened if the radioactivity of uranium had been discovered earlier. The element which was afterwards called uranium was discovered by Klaproth in 1789, more than a century ago, and, if he had put that substance near an electroscope he might have noticed that it discharged electricity, but in my opinion that would have been all. People would have said it was curious but would not have thought it of any consequence. No one would have asked how the effect was produced. It is characteristic of science that discoveries are rarely made except when people's minds are ready for them.

Now I hope you will allow me to give you an account of my personal acquaintance with the subject of radioactivity. When I entered the Cavendish Laboratory in 1895 I began work on the ionization of gases by X-rays. After reading the paper by Becquerel I was curious to know whether the ions produced by the radiation from uranium were of the same nature as those produced by X-rays, and in particular I was interested because Becquerel thought that his radiation was somehow intermediate between light and X-rays. I therefore proceeded to make a systematic examination of the radiation and I found that it was of two types, one which produced intense ionization and which was absorbed in a few centimetres of air, and the other which produced less intense ionization but was more penetrating. I called these α-rays and β-rays respectively, and when, in 1898, Villard discovered a still more penetrating type of radiation he called it γ-radiation.

In 1898 I went to McGill University, Montreal, and there I met R. B. Owens, the new Professor of Electrical Engineering, who had arrived at the same time as myself. Owens had a scholarship which required him to do some physical research, and he asked me whether I could suggest a problem which he might investigate to justify this scholarship. I suggested that he might become familiar with the use of an electroscope by studying thorium, the radioactivity of which had in the meantime been discovered by Schmidt. I assisted him with his experiments and we found some very queer effects. It appeared that the radioactive effect of thorium oxide would pass through dozens of sheets of paper put over the oxide but that it was stopped by the thinnest sheet of mica, as though something was being emitted which could diffuse through the pores of the paper. The fact that the apparatus was very sensitive to the effects of draughts supported this diffusion idea. We next did experiments in which air was drawn over the thorium oxide and then into an ionization chamber, and these showed that the activity could be transferred with the air. However, if the air current was stopped, the activity in the ionization chamber did not cease at once but gradually died away in an exponential manner. I gave the name of "thorium emanation" to this gaseous substance which could diffuse through paper, and could be carried away with the air and which preserved its activity for some time, decaying with a characteristic law.

I found that this emanation had the most peculiar property that when it was passed over bodies it made them radioactive. This appeared to be due to the deposit of a material substance, rather than to any activity induced in the bodies themselves under the action of the radiation, since the amount of the material deposited could be increased by applying an electric field. Many people at this time were obtaining capricious and peculiar results from materials placed near radioactive substances, and it seemed that these could probably all be explained by the presence of emanations of the type we had found in connection with thorium.

Before this explanation could be shown to be correct it was necessary to discover the exact nature of the emanation. This was very difficult, because the amount available was always very small. Soddy and I concluded, early on, that it must be one of the inert gases like helium, neon, and argon, since it was never possible to make it combine with any chemical substance. We were able to make a rough estimate of its molecular weight by comparing its rate of diffusion with that of other gases with known molecular weights. By using the property of discharging an electroscope as a measure of the amount of emanation present, we were able to measure these diffusion rates with very small quantities of emanation. We concluded that the atomic weight must be of the order of 100. We next tried to find whether the emanation was produced directly from the thorium, or from some intermediate product. Using chemical methods we were able to separate an intermediate substance, which we called thorium X, from which the emanation was produced.

About this time Ramsay showed that helium was present in most radioactive minerals, and that it represented another gaseous product of the transformations. Later on I was able to show that the helium was due to the accumulated α-particles.

Radium was not available in any quantity till 1903 or 1904, and most of what there was in the world was in the possession of the Curies, who had separated it by a long and arduous process from pitchblende. One of the first observations they made was that a quantity of radium weighing about 100 mg. kept itself above the temperature of the surrounding air, and they deduced that a gram of radium would emit heat at the rate of 100 calories per hour. This experiment created great excitement, because the idea of any substance keeping itself permanently at a temperature higher than its surroundings was repugnant to the old-fashioned physicists, and the prevailing idea became common that radium had a peculiar property of acting as a thermodynamic engine using heat from the air. I was firmly of the impression that the heating effect was a necessary consequence of the emission of the α- and β-particles and that it decreased with time in the same way as the activity. Later on we were able to classify the heating effects of radioactive bodies and to show that there was nothing obscure about the process. We were able to show that heat can be evolved in enormous quantities in these radioactive changes; when reckoned per unit mass of the material these quantities are millions of times greater than those given by chemical reactions, and we were able to show that this is a characteristic of all radioactive changes.

Now I would like to say a little about the experimental proofs of the nature of the α-rays. By various experiments and with the help of various collaborators I was able to show, by deflecting α-particles in magnetic fields, that these particles were helium atoms carrying two positive charges, and we were also able to determine their speed. About this time (1903 and 1904) Bragg and Kleeman made their very interesting and important analysis of the ionization curve of the α-rays, showing that the ionization varied along their path in a character-istic way. A curve showing the form of this variation is now known as a "Bragg curve".

Next I want to speak of two very important dis-coveries, the credit for which is due in a large measure to Prof. Soddy. I refer to the discovery of the displace-ment law, and the discovery of isotopes amongst the radioactive elements.

Soddy had been investigating the chemical properties of the radioactive substances, and he had noticed that there was often a simple relation between the positions in the periodic table of the original and the final elements involved in a radioactive disintegration. Before he could be sure of the generality of this result it was necessary to determine the chemical properties of all the known radioactive elements, not an easy matter, since many of them were only available in minute quantities. Similar work was being done by Prof. Hahn, and finally the broad generalization now known as the "displacement law" was made almost at the same time by Dr A. S. Russell, Prof. Fajans, and Soddy. This law stated simply that when a substance emitted an α-particle it moved two places down in the periodic table, and when it emitted a β-particle it moved one place up in the table. This was seen to be a consequence of the fact that an α-particle carries two positive charges and a β-particle one negative charge.

As regards isotopes the position was as follows. Many people had observed that there was an incredible difficulty, amounting almost to an impossibility, in separating certain radioactive bodies from one another. Soddy became very interested in this phenomenon and found there were some radioactive substances which he could not separate. These bodies were completely distinct and had characteristic radioactive properties, yet they could not be separated by chemical operations. He also pointed out that there was not enough room in the periodic table for the great group of radioactive elements, and he suggested that there were elements which from the chemical point of view were inseparable, but from the radioactive point of view showed different properties. Soddy called related elements of this kind isotopes, and that was the beginning of that great field of investigation which has owed so much to Dr Aston, and about which he will be telling you in his own lecture.

THE DEVELOPMENT OF THE THEORY OF ATOMIC STRUCTURE

In my lecture to-day I shall try to tell you very briefly something about the development of our ideas with regard to the constituent particles of which atoms are made, and the way in which these particles are combined to make up an atom.

One of the most important particles for our problem is the electron, and I shall try, very briefly, to show you first how our ideas about the electron have changed during the last forty years. It was in 1897 that the experiments, largely of our own leader, Sir J. J. Thomson, led to the conclusion that the so-called cathode rays of Crookes consisted of a stream of particles of minute mass travelling with very great speeds. I believe we are right in assigning a predominant part in that discovery to J. J. Thomson, for he was the first to deflect the particles in an electric as well as a magnetic field, and also the first to recognize that the electron must be a constituent of all atoms, and he it was who first devised methods of determining the number of electrons within an atom. These early experimenters found that the ratio of the charge to the mass of the electron was about one or two thousand times greater than that for hydrogen, the lightest known atom, and at the same time they showed that electrons in a vacuum tube may have very great speeds, approaching even that of light. Now the mass of the electron was not known, only the *ratio* of charge to mass, but all the indications were that the electron was very light and mobile, and that very interesting Scotsman, Sutherland, in Melbourne, suggested that this light electron might be nothing more nor less than a unit electrical charge in motion with no material mass associated with it. J. J. Thomson had shown in 1881 that a sphere of radius a carrying a charge e appeared to have an extra mass $\frac{2}{3}e^2/a$ corresponding to the fact that when it was set in motion energy had to be put into the electromagnetic field surrounding it. Sutherland pointed out that if the radius a were only supposed small enough there was no necessity to assume that the electron had any "ordinary" mass at all. For this to be true the radius would have to be about 2×10^{-13} cm. It was an attractive idea, and people set about trying to test its validity.

Theoreticians such as Abraham, Heaviside, and Searle here in Cambridge, tried to find out how the apparent mass due to the charge would vary with the velocity. Different investigators arrived at different results, owing to the fact that they made different assumptions to start with, but for moderately great speeds these results were roughly the same. All showed that the mass should increase with speed and should become infinite as the speed of light was approached. In the meantime small quantities of radium had become available, and as this emitted electrons travelling with velocities very close to that of light it was possible to make an experimental test of these theories. This Kaufmann did in 1902 and he got results which were in general agreement with all the theories, to the order of accuracy of the experiment.

These experiments attracted very considerable attention and led many people to the unjustified conclusion that, since the mass of the electron appeared to be entirely due to its charge, therefore all mass was nothing but a manifestation of electric charge. On this idea the mass of the hydrogen atom—1850 times that of the electron—was simply explained by supposing that the atom contained 1850 electrons. This stage, however, did not last long. In 1905 Einstein showed, from relativity ideas, that the mass of a body should change with its speed, and that it does not matter whether it is charged or uncharged, the change in mass is just the same. Every body, no matter what it consists of, must obey the Einstein law, and all experiments seem to show that this law is correct. Kaufmann's experiments agreed with the relativity results just as well as with the older electrical theories, so that it was no longer possible to suppose that the mass of the electron was entirely due to its charge. Since the only method of estimating the radius a of the electron was to assume that the mass was due entirely to the charge, and then use the expression given above, it is clear that once more there was no estimate of the size of the electron. It is probable that the radius is of the order 10^{-13} cm., and recently Prof. Born has evolved a theory which leads to a value of this order, but it is early yet to say whether that theory is correct.

We were quite happy for ten or fifteen years with the idea of the electron as a spherical distribution of charge, possibly together with some "ordinary" mass. In 1925, however, in order to explain some of the complications of the spectra of hydrogen and helium, Uhlenbeck and Gousmidt suggested that the electron also had a magnetic moment, and as they realized that a spinning spherical charge would have a moment of this kind they postulated a "spinning electron". Shortly afterwards, in 1930, Dirac developed a general theory in which relativity and wave-mechanics were combined, and he found that he could explain the fine structure in the spectra without postulating a special "spinning electron". At first it looked as though the idea of the "spinning electron" was not correct, but it appears now that Dirac has come to the conclusion that, on his theory, the electron must behave as though it had a magnetic moment, though there is no need to postulate this separately. It could not help behaving like that anyway.

It is next of interest to give some account of the determination of the electronic charge e, since this quantity is so intimately connected with the evaluation of atomic magnitudes. The first experiments were made by Townsend in the Cavendish while I was there in 1897. He found that a cloud condensed on hydrogen which had been produced by electrolysis and bubbled through water. This cloud was also found to be charged, and he determined the charge on each droplet in the following way. The weight of the whole cloud was found by precipitating it and weighing on a balance. The average weight of each drop was found by measuring the rate of fall of the cloud and using Stokes' law. Hence the number of drops was known. Since the total charge carried by the cloud could also be measured it was possible to find the charge on each drop. The method did not give a good value for the electronic charge, because many of the drops were multiply charged, but it is interesting because it included practically all the ideas which were later used in accurate measurements of the charge.

In 1908–13 J. J. Thomson used a method in which a cloud was produced by expansion, and its weight estimated from the known expansion ratio. Wilson applied an electric field so that the charged drops could be held stationary or driven up or down. In 1908 Geiger and I counted the number of α-particles emitted from a certain quantity of radium and then measured the total charge which they carried. We obtained a value $4 \cdot 65 \times 10^{-10}$ e.s.u., considerably greater than the value of $3 \cdot 4 \times 10^{-10}$ deduced by Thomson, but we did not think of our method as being at all accurate. In that connection Prof. Planck once told me an interesting story. When he first put forward his quantum theory of light, people were slow to believe it, partly because the theory required the electronic charge to be $4 \cdot 7 \times 10^{-10}$, whereas the accepted value was then $3 \cdot 4 \times 10^{-10}$. Planck himself was doubtful because of the discrepancy, but when Geiger and I announced the value 4.65×10^{-10} he began to be certain that his theory was correct.

The magnitude of the charge was, as you know, measured accurately by Millikan between 1910 and 1917. There is some doubt at the present day as to whether his result is as accurate as was originally believed, but I will not deal with that question here.

Now I come to a most interesting discovery of recent times. Many people had thought that in a properly constituted universe there ought to be a certain degree of symmetry, and where we had a negative electron we ought also to have a positive electron of the same small mass. Although this had often been looked for, it was not found until 1931, when Anderson, in California, was photographing the tracks of cosmic-ray particles as revealed in a Wilson cloud chamber. A strong magnetic field was applied to the chamber and he found that some of the tracks were curved in one direction and some in the other, showing that some represented negative particles and some positive. Other evidence showed that the masses of both were small and of the order of the electronic mass. Anderson got photographs showing these tracks only very rarely, but in 1933 Blackett and Occhialini, in the Cavendish Laboratory, developed a method by which the cosmic ray was made to trip the apparatus and "take its own photograph" so to speak. By this method it was possible to get many photographs of the tracks of positive electrons, or "positrons" as they are now called.

Blackett interpreted these results in terms of a theory developed in 1931 by Dirac. This theory had suggested that positive electrons might exist, but that their life would be very short since they would combine with the first negative electron they encountered and give rise to energy of radiation. In a sense Dirac had predicted the positive electron before it was discovered, but the prediction was well hidden in the theory. Theory and experiment both indicated that under suitable conditions radiation energy of very short wave-length, such as is present in the cosmic radiation, can disappear and give rise to a pair of electrons, one positive and one negative. This occurs most readily in the intense electric field surrounding a heavy nucleus, and is only possible if the quantum energy of the radiation is greater than one million electron volts, which is the equivalent of the mass of the electron-pair.

We now turn to consider the question of atomic structure. In 1895 Lennard made a famous experiment in which he passed electrons through a thin window in the discharge tube where they were made, and was able to observe them outside the tube. Since the electrons could penetrate the windows so easily he concluded that the atoms in the window must have a very open structure and have comparatively large spaces between them. He suggested that the atoms might contain spheres of positive electricity associated somehow with negative charges. A year or two later J. J. Thomson elaborated this idea and calculated how negative electrons would distribute themselves throughout a sphere of positive charge. He was able to explain in this way the fundamental nature of the periodic table.

Now I myself was very interested in the next stage, so I will give you it in some detail, and I would like to use this example to show how you often stumble upon facts by accident. In the early days I had observed the scattering of α-particles, and Dr Geiger in my laboratory had examined it in detail. He found, in thin pieces of heavy metal, that the scattering was usually small, of the order of one degree. One day Geiger came to me and said, "Don't you think that young Marsden, whom I am training in radioactive methods, ought to begin a small research?" Now I had thought that too, so I said, "Why not let him see if any α-particles can be scattered through a large angle?" I may tell you in confidence that I did not believe that they would be, since we knew that the α-particle was a very fast massive particle, with a great deal of energy, and you could show that if the scattering was due to the accumulated effect of a number of small scatterings the chance of an α-particle's being scattered backwards was very small. Then I remember two or three days later Geiger coming to me in great excitement and saying, "We have been able to get some of the α-particles coming backwards...". It was quite the most incredible event that has ever happened to me in my life. It was almost as incredible as if you fired a 15-inch shell at a piece of tissue paper and it came back and hit you. On consideration I realized that this scattering backwards must be the result of a single collision, and when I made calculations

I saw that it was impossible to get anything of that order of magnitude unless you took a system in which the greater part of the mass of the atom was concentrated in a minute nucleus. It was then that I had the idea of an atom with a minute massive centre carrying a charge. I worked out mathematically what laws the scattering should obey, and I found that the number of particles scattered through a given angle should be proportional to the thickness of the scattering foil, the square of the nuclear charge, and inversely proportional to the fourth power of the velocity. These deductions were later verified by Geiger and Marsden in a series of beautiful experiments.

Now let us consider what deductions could be made at that stage. By considering how close to the nucleus the α-particles could go, and yet be scattered normally, I could show that the size of the nucleus must be very small. I also estimated the magnitude of the charge and made it about a hundred times as great as the electronic charge e. It was not possible to make an accurate estimate, but general evidence indicated that the nucleus of hydrogen must have a charge e, helium $2e$, and so on. Geiger and Marsden examined the scattering in different elements and found that the amount of scattering varied as the square of the atomic weight. This result was rough but quite sufficient: it indicated that the charge on a nucleus was roughly proportional to the atomic weight.

At this time the idea that charge and atomic numbers were related was prevalent in our Laboratory, and it was then that Moseley began his famous experiments on X-rays. He showed that the X-ray spectra of elements varied regularly and uniformly from one element to the next, the spectra all being similar but shifted in frequency as we pass from element to element. Now, on the nuclear theory, the X-ray spectrum is presumably connected with the movement of electrons very close to the nucleus, and Moseley's experimental results led to the conclusion that the X-ray properties of the elements were dependent on the square of the whole number, which varied by unity from one element to the next. Moseley supposed that the atomic number represented the charge on the nucleus, and starting with aluminium 13, he was able to explain the X-ray properties of the elements up to gold, and the series was extended right up to uranium in 1932.

This theory at once showed which elements were missing in the periodic table, and where one ought to look to discover new elements. It was now clear that the atomic weight, which the chemist had previously supposed to be the important factor in the periodic table, must be replaced by the atomic number, and the properties of all the elements ought to be explicable in terms of whole numbers. The essential point of the identity of the atomic number and the nuclear charge was experimentally verified by Chadwick after the war.

This nuclear idea at once explained in a general way the existence of isotopes. The nuclear charge controls the arrangement of electrons and this arrangement in turn determines the chemical properties. We should therefore anticipate that isotopes should be bodies with the same nuclear charge but with different nuclear masses. As we know, this has been completely confirmed by Aston's later work.

Now we come to that question with which Niels Bohr's name is associated, "How are the electrons arranged in the outer atom?" Bohr's original quantum theory of spectra was one of the most revolutionary, I suppose, that was ever given to science, and I do not know of any theory that has been more successful. He was in Manchester at the time, and, being a firm believer in the nuclear structure of atoms as shown by experiments on scattering, he tried to see how he could arrange the electrons so as to give the known spectra of the atom. His success lay in bringing entirely new ideas into the theory. He imported into the picture the idea of the quantum of action, and he imported also the idea, foreign to classical physics, that an electron might circulate in an orbit round the nucleus without radiating. I was perfectly aware when I put forward the theory of the nuclear atom that according to classical theory the electrons ought to fall into the nucleus, but Bohr postulated that, for some unknown reason, they did not do so, and with this idea he was able, as you know, to give an explanation of the origin of spectra. He then passed from stage to stage, making certain reasonable assumptions, to work out the distribution of the electrons in all the atoms of the periodic table. There were many complications, since the distribution had to agree with the optical and the X-ray spectra of the elements, but in the end he was able to suggest an arrangement of electrons which showed the meaning of the periodic law.

As a result of later developments, largely influenced by Bohr himself, and modifications by Heisenberg, Schroedinger and Dirac, the whole mathematical theory has changed, and the idea of wave-mechanics has been introduced. Quite apart from these later developments I consider the work of Bohr to be one of the greatest triumphs of the human mind. To realize the significance of his work you have only to consider the incredible complexity of the spectra of the elements and to think that within ten years all the main features of these spectra had been understood, so that now the theory of optical spectra is believed to be so completely settled that many people consider it a dead subject, like sound was some years ago.

We must now pass to the development of later ideas on the structure of the nucleus itself. In 1919 I showed that when light atoms were bombarded by α-particles they could be broken up with the emission of a proton, or hydrogen nucleus. We therefore presumed that a proton must be one of the units of which the nuclei of other atoms were composed, and the theoreticians set to work to try and explain the properties of nuclei by combination of protons and negative electrons. It is, however, very difficult to combine the slow and ponderous proton with the light and lively electron in such a confined space as a nucleus, and it was not until Chadwick brought to light the existence of an uncharged particle, the neutron, that the problem appeared theoretically soluble. It was then possible to suppose that the nuclei of all atoms consisted of a combination of protons and neutrons, so that, for example, oxygen with charge 8 and mass 16 had 8 protons and 8 neutrons. This was a very simple idea, and the valuable point was that the constituent particles had similar mass. But what are we going to do about the fact that a negative electron often comes out of a nucleus in radioactive changes, and that a positive electron comes out in certain artificial transmutations? In answer to this the theoretician suggests that, in the confined space of the nucleus, where the force between the particles is enormous, protons may change into neutrons and *vice versa*. For example, if a neutron lost a negative electron it would pass into a proton, and if a proton lost a positive electron it would become a neutron, so that in the first case a negative particle, and in the second a positive particle, could be emitted. The electrons and positrons do not exist free in the nucleus, they are bound to the neutron or the proton as the case may be, and they are only released under certain conditions of great energy change within the nucleus.

I have tried to give you a general idea of the way in which we started to investigate these matters forty years ago, and of the way in which the ideas have developed stage by stage. I have also tried to show you that it is not in the nature of things for any one man to make a sudden violent discovery; science goes step by step, and every man depends on the work of his predecessors. When you hear of a sudden unexpected discovery—a

bolt from the blue as it were—you can always be sure
that it has grown up by the influence of one man on
another, and it is this mutual influence which makes
the enormous possibility of scientific advance. Scientists
are not dependent on the ideas of a single man, but on
the combined wisdom of thousands of men, all thinking
of the same problem, and each doing his little bit to add
to the great structure of knowledge which is gradually
being erected.

IV. FORTY YEARS OF CRYSTAL PHYSICS

by

W. L. BRAGG

*Cavendish Professor of Experimental
Physics, Cambridge*

FORTY YEARS OF CRYSTAL PHYSICS

I HAVE been asked to deal in this lecture with the development of crystal physics during the last forty years. Any such treatment comes under three headings —the history of its rapid rise to an important branch of science, the position reached at the present time, and the prediction of the probable lines of advance in the future. The title "Crystal Physics" is perhaps rather misleading in that it lays too much emphasis on the crystalline nature of the substances which are studied. "Physics of Atomic Arrangement" is a clumsy title, but one which corresponds more closely to the nature of the subject. This branch of science deals on the one hand with the physical methods used for discovering the arrangement of the atoms in matter, and on the other hand with the explanation of the properties of matter in terms of this arrangement. The atoms are arranged in a regular way in crystals, hence crystals afford the easiest approach to the problems of analysis. They are only a means towards an end, however. We are not so much interested in the crystalline state in itself, as in the information afforded by the positions taken up by the atoms relatively to each other. In organic crystals, for example, we are interested in the *molecule*. The molecules pack together in a regular crystalline manner. Measurements made on such crystals yield a picture of the molecule with data about interatomic distances and bond-angles, and this is the final goal of the investigation. We must include methods for investigating the structure of amorphous bodies such as glass, of liquids, and of gases, which are identical in principle with those used for crystals. There are cases when the study of the crystalline state is an essential part of the problem. Minerals are bodies of this kind, for the majority of them can only exist as crystals. It is for this reason that mineralogy and crystallography have had so long and honourable an association. The crystalline state is essential to a mineral because only in that way can the extremely low potential energy be attained which has preserved the substance unchanged throughout geological ages. It must be realized, however, that whereas crystallography has historically been a relatively unimportant though interesting branch of science studied by a few enthusiasts, what is now termed "Crystal Physics" is a study of infinitely wider scope. It has a bearing on all other sciences and is one of the most important branches of modern physics. We may distinguish two territories in physics, whose common frontier is the outside of the atom. The first embraces investigations into the interior of the atom and nucleus. The second embraces investigations which start with the atom as the elementary particle and study the bodies formed by packing the atoms together. Though the former leads to ever more fundamental physical concepts, it is the latter which is having such a profound influence on other branches of science. Chemistry, biochemistry, metallurgy, mineralogy, are having their outlook modified to an extent which is perhaps hardly realized.

The reason for this is that X-ray analysis has provided us with a means of extending optical investigation to the scale of atomic structure. The revolution which has been effected is comparable to that produced by the invention of the microscope. We are effectively able to see the atoms, for the X-ray microscope can be applied to any form of matter and its atomic pattern studied. Devotees of this new science are sometimes idly asked what will happen when all common crystals have been examined —whether this task will not then come to an end. Hooke might equally well have been asked the same question when he first described the various structures he observed under the microscope, yet the microscope can hardly be said to have lost its scientific importance in the centuries subsequent to Hooke's work.

The year 1895 is a convenient starting-point because at that time the geometrical theory of crystal structure was developed in a complete way, by the independent work of Schoenflies, Fedorow, and Barlow. Three-dimensional patterns may be distinguished by their schemes of symmetry. These authors showed that patterns of any kind, not necessarily crystal structures, belong to one or other of 230 types. Each type has its own characteristic array of axes, planes, and centres of symmetry. Though it was not possible at that time to deduce the scheme of symmetry or "space-group" of any actual crystal, yet the geometrical theory of space-groups accounted fully for the external symmetry displayed by crystals, showing why only certain types of external symmetry (thirty-two in all) would be possessed by any solid with a regular atomic pattern. It was also about this time that exceedingly interesting speculations about the atomic arrangement in crystals were made by Pope and Barlow in their valency-volume theory. The postulate that atoms in a crystal occupied a volume proportional to their valency has proved fallacious, but in developing their theory they pictured possible atomic patterns many of which have since been proved actually to exist. Their models of cubic and hexagonal close-packing of equal spheres, and of simple structures of the potassium chloride and caesium chloride type, gave a reality to crystal patterns and were a source of inspiration to early workers in X-ray analysis.

Laue discovered X-ray diffraction by crystals in 1912. An account of the discovery will be found in an article by Friedrich in *Naturwissenschaften* for 1922. Laue was led to his discovery by speculating on the probable interaction between very short electromagnetic waves, such as X-rays were supposed to be, and the regular array of atoms in a crystal. As in the case of many other epoch-making discoveries, it seems extraordinary to us now that the effect had not been previously discovered accidentally. A narrow beam of X-rays, a fragment of crystalline matter in their path, and a photographic plate which received the scattered rays, at once revealed the existence of diffracted beams.

I was very interested in Laue's discovery because at that time my father was an advocate of the view that

X-rays were corpuscular and not undulatory in nature.
He was led to this view by a series of experiments from
which he deduced that the ionization of a gas traversed
by X-rays was not a primary effect of the rays, but due
to β-rays excited in the gas by the X-rays. This conclusion
we now know to be correct (the Wilson cloud chamber
had not then been invented), but the quantum theory
has reconciled the undulatory and particle aspects of
electromagnetic radiation. In 1913 I undertook experiments to see whether Laue's apparent diffraction effects
were due to particles travelling down avenues in the
crystal pattern. I was of course soon convinced that
Laue had correctly ascribed the pattern of beams to
diffraction by the crystal grating. I found, however, that
it was possible to explain certain peculiar features of the
pattern, not by complexities in the X-ray spectrum as
Laue had done, but by the nature of the crystal structure,
and so was led to the idea of *using X-ray diffraction to
analyse the atomic arrangement in crystals*. The diffraction
effects which Laue had observed with the crystal zincblende, ZnS, were for instance explained by assuming
that the crystal had a face-centred cubic pattern and not
the simpler pattern of scattering units at cube corners
only. Prof. Pope took a great interest in this work as it
touched so nearly on his theories of crystal structure,
and he suggested an investigation of the crystals of
potassium and sodium chlorides. The diffraction patterns given by these crystals proved to be very simple,
and I was able to show that their structure was a chess-
board pattern of alternate K (or Na) atoms and Cl

atoms. In the course of this work I treated the diffraction of X-rays as a *reflection* of the rays by the sheets of atoms in the crystal according to the law

$$n\lambda = 2d \sin \theta,$$

where λ is the wave-length of the X-rays, n an integer, d the spacing of the crystal planes and θ the glancing angle at which the X-rays fall on the planes. By using a cleavage face of a crystal, parallel to a set of planes, X-rays were regularly reflected as if by a mirror. This experiment aroused considerable interest at the time, though it was really only an alternative way of interpreting Laue's original analysis.

My father thereupon constructed an apparatus for examining the reflected beam, the first *X-ray spectrometer*, in order to test whether this beam had the properties of X-rays. The beam was measured by allowing it to enter an ionization chamber. The result of his experiments was the discovery of *X-ray spectra*. He showed that the rays from a platinum anticathode contained monochromatic components, a line spectrum, as well as continuous "white" radiation.

This discovery had two results. In the first place, it provided a much more powerful method of crystal analysis. I had used the Laue method for analysing KCl and NaCl, in which white radiation was reflected simultaneously from all the planes of a stationary crystal in the path of the X-ray beam. The planes selected the radiation of correct wave-length for reflection according to the Bragg law. On the other hand, when using monochromatic radiation, each set of crystal planes was examined in turn, and a measure of the angle of reflection gave the spacing of the planes directly. Comparison of the intensities of different orders indicated the structure of the planes. At this stage my father and I joined forces, and I was privileged to work with the X-ray spectrometer. The structures of a number of simple crystals, zincblende, diamond, fluorspar, iron pyrites, and calcite were rapidly determined.

In the second place, the ionization spectrometer was the start of X-ray spectroscopy. My father determined the spectra of X-rays from platinum, osmium, indium, palladium, rhodium, copper, and nickel anticathodes. He discovered the K_{α} and K_{β} lines, and the L_{α}, L_{β}, L_{γ} lines. He showed that their energy quanta $h\nu$, according to Planck's relationship, tallied with the cathode-ray energies necessary to excite the K and L radiation which had been measured by Whiddington. He showed that the frequency of the K lines was roughly proportional to the square of the atomic weight. The foundations were laid, in fact, for the subsequent work of Moseley, who by measuring a series of spectra was led to his brilliant generalization relating frequency to *atomic number*. Absorption edges were also discovered and measured by the X-ray spectrometer.

Two important theoretical contributions were made in 1913 and 1914. Darwin laid the foundations of the theory of X-ray reflection in two exhaustive papers in the *Philosophical Magazine*. Debye calculated the effect of thermal movements in weakening the X-ray reflection. It is remarkable to what an extent all subsequent theoretical and experimental advances were foreshadowed in the first two years after Laue's discovery of diffraction.

We may perhaps end the account of the early history at this point and describe the course of events in the next twenty years.

A striking feature is the extension of X-ray analysis to increasingly complex atomic patterns. The first crystals to be analysed were of exceedingly simple type. Atoms were either fixed in position by the symmetry (structures of no parameters) or one or two parameters were sufficient to determine the atomic arrangement. For ten years (up to 1923) it was considered practically impossible to analyse crystals in which the atoms were in "general" positions, or crystals of lower symmetry than cubic, tetragonal, or hexagonal. The overcoming of these limitations was effected as follows. In the first place, systematic methods were developed for deducing the space-group or symmetry scheme of a crystal. Niggli was the first (1919) to draw up *tables for space-group determinations*, followed by Wyckoff, and Astbury and Yardley. Then it was shown by workers in my school at Manchester that *absolute measurements of intensity* of reflection enabled us to handle crystals of low symmetry with many parameters. Aragonite, barytes, phenacite were amongst the first to be analysed, followed by a number of complex silicates. This work was criticized at the time as being too speculative, but experience has shown that it was really on safe lines. Next, as knowledge of more complex structures increased, it was seen that certain guiding principles could be used in the search for structures which would explain the X-ray results. Atoms pack into the crystal structure as if they occupied characteristic approximately constant volumes. Acid radicles have characteristic forms. Pauling's rules (1929) for the association of ions in inorganic crystals are invaluable in structure determination. The combination of these new principles led to a more and more confident handling of highly complex structures, and it may now be claimed that no inorganic crystal is too difficult to analyse; it is sure to yield to a determined attack.

The analysis of organic crystals was initiated by W. H. Bragg in 1921 (naphthalene and anthracene). The method of attack was here on quite different lines. The structures assigned by organic chemistry to the molecules were assumed to be correct, and a comparison of the naphthalene and anthracene unit cells showed how the molecules lay in these cells. Shearer, Piper, and de Broglie and Friedel, simultaneously analysed long-chain aliphatic compounds in a similar way. This pioneer work has now developed into a series of investigations of very complex organic molecules. Robertson's work on aromatic compounds, and Muller's work on hydrocarbons and related long-chain compounds, are outstanding examples.

Westgren initiated the study of alloy structures by X-rays, of which further mention will be made below.

Herzog, Jancke, and Polanyi made the first attack on a biochemical substance, cellulose, in 1920, and its probable detailed structure has been analysed by Mark and Andress. Katz in 1925 discovered that stretched rubber gives a crystalline pattern and this has led to a new conception of the nature of rubber-like bodies. Astbury, in 1931, attacked what is probably the most complex and difficult substance to be examined by X-rays, the structure of keratin. The recent work of Bernal on proteins carries on the story.

The attack on all these problems has been made possible by the development of new technical methods of X-ray analysis. The *powder method* devised in 1916 by Debye and Scherrer and independently by Hull in 1917 enables microcrystalline bodies to be studied. The *rotation photograph* was first used by Schiebold in 1918. The Weissenberg camera enables the diffracted beam to be classified in a simple way. Various methods of microphotometry have raised to a high degree of accuracy the measurement of the strength of the diffracted beams.

We may at this stage review the effect of these discoveries on other sciences. The position as regards chemistry is extremely interesting. To put it briefly, the discoveries of X-ray analysis have shown that the conclusions about molecular structure arrived at by organic chemistry were correct to an extent which must excite our enthusiastic admiration, but, on the other hand, have shown that the fundamental ideas of inorganic chemistry need a very thorough revision. Organic chemistry is properly based on the conception of the molecule, an entity built of a collection of atoms with a definite stereochemical relationship to each other. The nature of the molecule is determined by the relative positions of the atoms, and these positions have been deduced from chemical considerations. X-ray analysis has done little more than make these pictures of the molecule more exact, by measuring interatomic distances and bond-angles. In practically every case, even the most complex, the structure assigned to the molecule by the chemist has been confirmed as correct.

On the other hand, we now realize that inorganic chemistry has adopted this same idea of a molecule, and tried to apply it generally, with disastrous results. Typical inorganic compounds in the solid state are essentially continuous structures. The very first crystals to be analysed, KCl and NaCl, showed the fallacy of the idea of KCl and NaCl molecules. They are structures of alternate positive and negative ions, with no indication of any association into pairs. It is extraordinary how hard the idea of the inorganic molecule has died; many of us will recall how earnestly we were begged to find some slight indication of a pairing into KCl molecules in our X-ray investigations. A striking example of the way in which confusion has arisen through the attempt to think of inorganic compounds as groups of molecules is to be seen in the silicates. These are essentially composed of continuous silicon-oxygen skeletons enclosing positive ions. Regarded in this way, their relationships to each other and their great variations in composition are easily understood, and the classification of natural silicates is neat and orderly. Thought of as "molecules" with definite "formulae", as "solid solutions", as "salts of silicic acids" they become a bewildering maze.

Pauling's rules for inorganic compounds have shown us how we must interpret valency in such compounds. It has a stereochemical significance for inorganic compounds as for organic compounds, but one of a different kind. It is not sufficient merely to balance total positive against total negative in composition. It must also be geometrically possible so to arrange the atoms that local balancing of electric charge occurs. This cardinal law of inorganic valency is the direct result of the determination of atomic arrangement.

Another science on which the new knowledge has important bearings is metallurgy. Metals and alloys are crystalline. The phases which appear in the metallurgist's equilibrium diagram have each a distinctive crystalline pattern. X-ray analysis has provided a powerful method of checking and extending the equilibrium diagrams of binary alloys determined by the customary metallurgical methods. It is already clear that it is going to be of vital importance in determining ternary and more complex equilibrium diagrams. What is more, the knowledge of atomic arrangement has provided a foundation for the calculations of the theoretical physicist. We are on the eve of a new era in metallurgy; it should be possible to predict the result of alloying metals together, instead of being obliged to proceed in an empirical way.

If we try to predict the most interesting lines of future research, the establishment of a theory of alloy structure is certainly one of these. In organic chemistry, stereochemistry has already reached so advanced a stage that so far X-ray investigations have done little more than confirm and make more definite what is already known. X-ray analysis is nearing the point, however, where it may be expected to provide new knowledge of highly complex organic molecules whose structure has not yet been established on purely chemical grounds. In biochemistry there are immense possibilities. Analysis is extremely difficult, owing to the high complexity of the compounds and the imperfect nature of their organized arrangement, but the work on cellulose, keratin, and some proteins has shown what may be done. Another field, which has not yet been mentioned, is the X-ray study of the liquid and amorphous state. The work of Debye and Prins on liquids, of Warren and Zachariasen on glasses, has already given rise to a host of new ideas. The treatment of the structure of water by Fowler and Bernal is an indication of the fruitful co-operation of X-ray analysis and theoretical physics in this field. The present rate of progress is determined, not so much by the lack of problems to investigate or the limited power of X-ray analysis, as by the restricted number of investigators who have had a training in the technique of the new science, and by the time it naturally takes for its scientific and technical importance to become widely appreciated.

V. FORTY YEARS OF ATOMIC THEORY

by

F. W. ASTON
Fellow of Trinity College,
Cambridge

ATOMIC THEORY

I SHOULD like to state first that I am not responsible for the title of the lecture "Atomic Theory", but fortunately a good deal of that has been covered by Lord Rutherford in the earlier lectures, so I shall confine myself to that part of the subject dealing with elements and atomic weights.

When I started to learn chemistry in the early 'nineties my teachers were very confident when they spoke of elements and atomic weights. They had no doubt what they meant when they told me that the atomic weight of oxygen was 16, chlorine 35·5, magnesium 24·3 and hydrogen 1·008. The reason for this confidence arose from complete trust in Dalton's atomic theory of 1803. Dalton was a Manchester chemist, and he had put forward a theory that contained the famous postulate that atoms of the same elements were similar to one another and equal in weight. Shortly after that an Edinburgh physician made the suggestion that all atoms were made of the same primordial atoms of a substance which he called "protyle" and which he endeavoured to identify with hydrogen. The physician was Dr Prout, and he had said that the combining weights of all the elements should be whole numbers, but when the chemists examined these they found it was quite impossible that both theories should be right. The combining weights of the elements were found to be fractional, and they had to drop one or the other. They chose to work with the sound working hypothesis of Manchester, rather than with the more philosophical speculations of Edinburgh. An illustration of "what Manchester thinks to-day the world will think to-morrow"; it went on thinking so for something like a hundred years. During that time atomic weights were determined with greater and greater precision, and that important pioneer, Stas, did wonderful work in discovering the accurate atomic weights of chlorine, hydrogen and other elements. But in 1886 Crookes suggested that it was just possible that Dalton's postulate might not be true, and in his presidential address to the British Association at Birmingham said:

I conceive, therefore, that when we say the atomic weight of, for instance, calcium is 40, we really express the fact that, while the majority of calcium atoms have an actual atomic weight of 40, there are not a few which are represented by 39 or 41, a less number by 38 or 42, and so on.

Later, he developed this idea in connection with his pioneer work on the rare earths. He called the components "meta-elements", but unfortunately for his reputation as a prophet the experimental results on which his idea was founded were later proved to be fallacious, and Dalton's postulate was reinstated as an article of scientific faith more firmly than ever.

Dalton's postulate cannot be tested in general by chemical methods, for the smallest quantity of a substance of use in chemical operations contains countless myriads of atoms. I propose with the model I have in front of me to give you some idea of the extreme smallness of the atoms when we reach the stage at which further division will alter their properties, and they remain the atoms of the substance no longer. The substance I shall take for my example is lead. You may assume that this cube for the purposes of the lecture is made of lead. It is a 1 dcm. cube. The section will be made by means of the infinitely sharp knife in three dimensions in such a manner that the first cube formed is half the linear dimensions, and one-eighth the volume of the original cube. I will now repeat the operation in exactly the same way, and we reach what I shall call the second cube of the series; again repeating the operation with the model, I shall reach the third cube of the series. What I want you to notice is the extreme rapidity with which that series diminishes. Each time you get only half the linear dimensions, and one-eighth the cubical dimensions. The question is how long can we go on repeating this operation. Well, I cannot go on very far in actual practice with models, because the results would become invisible; but I can carry on the series to an indefinite degree by means of lantern slides. In Fig. 1 are shown the eleventh to the fifteenth cubes of the series, and to compare their sizes you have a few familiar objects drawn to scale. I may say in this series you reach the limits of accuracy of several means of analysis. The chemical balance will fail at the ninth cube, which does not figure on this at all. The quartz micro-balance will fail at the fourteenth, though it is capable of detecting one-millionth of a milligram. Spectrum analysis fails at the fifteenth; but the surprising thing is that with the ordinary microscope one can still see objects smaller than that. That is to say, in the detection of minute particles of matter spectrum analysis is not so sensitive

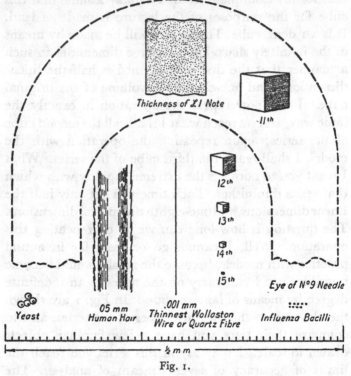

Thickness of £1 Note

-11ᵗʰ

12ᵗʰ

13ᵗʰ

14ᵗʰ

15ᵗʰ   Eye of N°9 Needle

Yeast   05 mm   .001 mm   Influenza Bacilli
             Human Hair   Thinnest Wollaston
                          Wire or Quartz Fibre

½ m m

Fig. 1.

as the ordinary microscope. Fig. 2 shows the seventeenth, eighteenth, nineteenth, twentieth, and twenty-first cubes of the series. Here there is some difficulty in finding the familiar objects. You see the wave-length of cadmium red light represented symbolically is very much bigger

than the oil film above it, and that is the reason the oil
film shows its colours. It is obvious that, although we
have not reached the atom at this stage, we are never

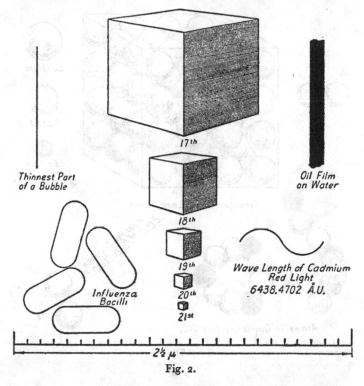

Fig. 2.

going to see it, because this wave-length of light is
enormously greater than the atom, so that you can
never see this with the eye. Fig. 3 shows the twenty-
sixth cube, and you see that two more operations will
reduce it to the single atom, so that this series of sections

can be carried on with lead 28 times before the atom of lead is reached. There are no familiar objects to compare on this slide, but up on the right you see drawn to scale

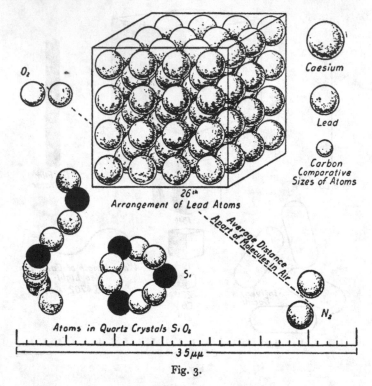

Fig. 3.

spheres representing the largest atom known, that of caesium, the smallest, that of carbon, and one of lead intermediate between these two. There are also represented on the same scale two molecules of air; a molecule of nitrogen, and a molecule of oxygen at their average distance apart in the air we breathe. There are also representations of the curious spiral form of the atoms of silicon and oxygen in quartz, which gives some indication for the reason of rotation of polarized light. Our definite knowledge of atoms, their size and position, and so on, is obtained almost entirely from the work of the Braggs and others on crystal analysis by X-rays. To give you some idea of the numbers of these atoms is difficult, as the numbers are so colossal. If the atoms in the original decimetre cube of lead were all put into a chain side by side the same distance apart as they are in the normal lead, the strings of atoms so formed would reach over six million million miles. A better idea is given by the fact that suppose you make a hole in an ordinary evacuated electric light bulb and allow the air molecules to pass in at the rate of 1,000,000 a second, the bulb will become full of air in approximately 100,000,000 years. Perhaps the most impressive illustration of all is to suppose that you could label the molecules in a tumbler of water. Suppose one was able, by some means, to do this in such a way that you would know them again, and you took the tumbler of water and threw it anywhere you please on the earth, and went away from the earth for a few million years while all the water on the earth, the oceans, rivers, lakes and clouds had had time to mix up perfectly. Now supposing that perfect mixing had taken place, you come back to earth and draw a similar tumbler of water from the nearest tap, how many of those marked molecules would you expect to find in it? Well, the answer is 2000. There are 2000 times more molecules in a tumbler of water than there are tumblers of water in the whole earth.

Dalton's postulate can be attacked experimentally by two entirely different methods. Physically by the development of methods by which the weights of individual atoms can be compared, and chemically by showing that it was possible to have samples of the same element with different atomic weights. The development of these two lines of attack took place about the same time, early in the twentieth century, and since the second and less direct scored the first success this will be considered first.

It was a direct outcome of the discovery of radioactivity in which the effects of individual atoms, as opposed to those of vast multitudes, were observed for the first time. Chemists could examine elements in the actual process of the making. In 1906 Boltwood observed that his newly discovered element ionium was so similar to thorium that if, by chance, their salts became mixed it was impossible to separate them by any chemical process. Other chemical identities among the products of radioactivity were soon observed and the most painstaking and delicate methods failed to effect or detect the slightest separation.

Discussing these, Soddy, in 1910, boldly stated:

These regularities may prove to be the beginning of some embracing generalization, which will throw light, not only on radioactive processes, but on elements in general and the Periodic Law.... Chemical homogeneity is no longer a guarantee that any supposed element is not a mixture of several of different atomic weights, or that any atomic weight is not merely a mean number.

The generalization underlying his views was the law connecting radioactivity and chemical change, in the discovery and enunciation of which he played so prominent a part. This law asserts that a radioactive element when it loses an alpha particle goes back two places in the periodic table; when it loses a beta particle it goes forward one place. It follows that by the loss of one alpha particle followed by two beta particles, the atom, though weighing four units less, will have regained its nuclear charge and returned to its original place.

Such changes result in bodies to which Soddy applied the following words:

The same algebraic sum of the positive and negative charges in the nucleus when the arithmetical sum is different gives what I call "isotopes" or "isotopic elements" because they occupy the same place in the periodic table. They are chemically identical, and save only as regards the relatively few physical properties which depend upon atomic mass directly, physically identical also.

This theory of Isotopes received the strongest criticism from all sides; it seemed so completely against the generally accepted facts. Particularly the idea that atoms of different weights could have identical spectra was extremely repulsive to orthodox physicists. Fortunately it was possible to put these revolutionary views to an experimental test in the case of one element—lead, the final inactive product of the thorium and uranium transformations. Uranium of atomic weight 238 loses eight alpha particles to become lead of atomic weight 206, while thorium of mass 232 loses six to become lead of atomic weight 208. Soddy maintained that the lead found in uranium minerals should be lighter, and that in thorium minerals heavier, than ordinary lead of atomic weight 207·2, and by 1914 had satisfied himself that this actually was so.

We will now look at the subject from the other and more general point of view, namely the measurement of the masses of the individual atoms. In order to weigh an atom we must give it a charge of electricity. This is most conveniently done by the electric discharge through gas at low pressure. In the intense field in front of the cathode of the discharge tube the atoms are broken up or "ionized". The negatively charged parts fly away from the cathode, forming cathode rays. These are electrons or atoms of the negative electricity, and are the same whatever the elements in the tube. There are also positive rays which travel towards the cathode. These will be the atoms which have had one or more electrons knocked off them and remain with a positive charge, and *these* will be the atoms of the gas you put into the tube. Owing to the very high field in front of the cathode, they shoot right through it, if a hole is provided, and cause a glow in the gas.

It was by this glow that they were first discovered in 1886 by Goldstein, who called them "canalstrahlen". It was more than twenty years before they were successfully analysed by Sir J. J. Thomson, who called them "positive rays" because they carried a positive charge of electricity. In his well-known parabola method of analysis the rays, generated by means of an electric discharge, after reaching the surface of the cathode enter a long and very fine metal tube. By this means a narrow beam of rays is produced which is subjected to deflection by electric and magnetic fields and finally falls upon a screen of fluorescent material or a photographic plate. The fields are arranged so that the two deflections are at right angles to each other. Under these conditions particles having the same mass but different velocities will strike the target on a parabola, and the position of this parabola will depend upon the mass. When this method of weighing atoms was used, all the results seemed at first to support Dalton's postulate; indeed the appearance on a sensitive screen of a clear-cut parabolic streak, caused by the impact of the atoms of hydrogen, was the first experimental proof that it was in any sense true of any element. Previously it had been purely an article of scientific faith. Hydrogen, carbon, nitrogen, and oxygen, present either as atoms or molecules, gave parabolas in the positions expected, and it was only when the rare gas neon was examined that an anomaly was observed. Neon, however pure, always gave two parabolas, a strong one at 20 and a weak one at 22. Referring to the latter in January 1913, Sir J. J. Thomson said:

The origin of this line presents many points of interest; there are no known gaseous compounds of any of the recognized elements which have this molecular weight. Again ,if we accept Mendeléev's Periodic Law, there is no room for a new element with this atomic weight.... There is, however, the possibility that we may be interpreting Mendeléev's law too rigidly, and that in the neighbourhood of the atomic weight of neon there may be a group of two or more elements with similar properties, just as in another part of the table we have the group iron, nickel and cobalt.

It was my privilege to be associated with him in this work, and as his attention was fully occupied with the investigation of a parabola of mass 3—now known to be triatomic hydrogen—it fell to my lot to search for a proof that neon was not homogeneous. This I endeavoured to do by partial separation of its hypothetical constituents, using as a test its density measured by a quartz micro-balance specially designed for the purpose. The first method, that of fractional distillation from charcoal cooled with liquid air, failed, as we now know was inevitable. The second, diffusion through pipeclay, though extremely tedious, had more success and I was able to announce in 1913 that, after thousands of operations, a definite change of density, amounting to about 0·7 per cent, had been achieved. Further data from positive rays were obtained, and, when the war stopped work, there were several lines of reasoning indicating that neon consisted of two bodies of different mass, and that the behaviour of these was exactly that predicted by Soddy for isotopes, but none of these was sufficiently strong to carry conviction on so important a conclusion.

During the war Soddy's prediction concerning the atomic weights of leads from uranium and thorium minerals had been triumphantly vindicated by some of his most severe critics, the experts in chemical atomic weights, and when work was started again, although

I continued for a time to experiment on separation by diffusion by means of an automatic apparatus, I realized that the most satisfactory proof of the existence of isotopes among the elements in general was only to be obtained by much more accurate analysis of positive rays. This was done by means of a sequence of electric and magnetic fields which gave focused images of fine collimating slits, thus forming a spectrum dependent upon mass alone. This I called a "mass-spectrograph". It had a resolving power of about 1 in 130 and an accuracy of mass measurement of 1 in 1000. This was ample to prove in 1919 that neon consisted, beyond doubt, of isotopes 20 and 22, and that its atomic weight 20·2 was the result of these being present in the ratio of about 9 to 1. Chlorine was found to contain 35 and 37, and bromine, of atomic weight almost exactly 80, and hence expected to be simple, gave two equally intense lines 79 and 81. Other elements were shown to be much more complex. Krypton, the first of these, had six isotopes, 78, 80, 82, 83, 84, 86; xenon and tin even more. Of the greatest theoretical importance was the fact that the weights of the atoms of all the elements measured, with the exception of hydrogen, were whole numbers to the accuracy of measurement. This "whole number rule" enabled the simple view to be taken that atoms were built of two units, protons and electrons, all the former and about half the latter being bound together to form the nucleus.

The mass of the hydrogen atom was determined by a special method and proved to be nearly 1 per cent greater than a whole number. This measurement made by means of the mass-spectograph in 1920 was of far-reaching significance because it proved quite definitely the possibility of sub-atomic energy. It is reasonably certain that the electrical particles forming four atoms of hydrogen are precisely the same as those forming one atom of helium, so that if we were able to transmute one into the other nearly 1 per cent of mass would be annihilated. On the relativity equivalence of mass and energy now experimentally proved, the quantity of energy liberated would be prodigious. Thus to change the hydrogen in a glass of water into helium would release enough energy to drive the "Queen Mary" across the Atlantic and back at full speed.

The discovery of isotopes in the elements generally made a very great change in the significance of atomic weight. I well remember interviewing Sir William Pope on this matter, and he suggested that in a few years time we should be making tons of chlorine 35 and tons of chlorine 37. Thinking of my recent experience with neon I said that I did not think this was at all likely, and so it was decided, rightly or wrongly, that the word "element" should be left undisturbed to be used as it always had been. This decision has been justified, for although sixteen years have elapsed only two elements have been separated into their component isotopes at all completely, in reasonable quantities, the one, hydrogen, is entirely exceptional, and the other, neon, has no chemical significance.

Although the change from the point of view of the

practical chemist is so small, to the philosopher it is profound, as is well illustrated by two quotations I will read. One is from Stas (1860–65):

I have arrived at the absolute conviction, the complete certainty, so far as it is possible for a human being to attain to certainty in such matters, that the law of Prout is nothing but an illusion, a mere speculation definitely contradicted by experience.

The other is from Soddy (1932):

After many vicissitudes and the most convincing apparent disproofs, the hypothesis thrown out so lightly by Prout, an Edinburgh physician, in 1815, has, a century later, become the corner-stone of modern theories of the structure of atoms. There is something surely akin to if not transcending tragedy in the fate that has overtaken the life work of that distinguished galaxy of nineteenth century chemists, rightly revered by their contemporaries as representing the crown and perfection of accurate scientific measurement. Their hard won results, for the moment at least, appear as of as little interest and significance as the determination of the average weight of a collection of bottles, some of them full and some of them more or less empty.

Although the interpretation of mass-spectra was often far from simple owing to the difficulty of distinguishing between lines due to compound molecules and those representing true atomic mass-numbers the analysis of the more suitable elements advanced rapidly. Dempster at Chicago discovered the isotopes of magnesium, calcium, and zinc by means of an instrument of his own design with semicircular magnetic focusing. By 1925, when I replaced my first mass-spectograph, now in the Science Museum, South Kensington, with one of higher resolving power, information on the isotopic constitution of more than half the elements had already been obtained. The new instrument was designed primarily for measuring the minute variations of the masses of atoms from the whole number rule, and had a resolving power ample for the heaviest elements. By its means many new isotopes were discovered.

The difficulty of obtaining the necessary rays for analysis varies enormously from element to element. Two main devices are employed: the ordinary gas discharge which requires the element to be volatile or form suitable volatile compounds; and the anode ray discharge, in which the halide or other compound of the element is treated as the anode in a discharge at low pressure. The inert gases are particularly suitable to the first method, the alkali metals to the second, other groups of elements being intermediate. Our knowledge of the mechanism of the discharge in both methods is far from complete, so that working with them is still rather an art than a science. The element of luck has played an important part in cases where the properties of the materials are unfamiliar and unfavourable to the conditions of the discharge. The technique of anode rays. is capricious, but when successful, yields spectra almost free from the lines of compound molecules, and is for this reason particularly suitable for the identification of new isotopes. I was able to apply it to my second mass-spectograph in the analysis of the large group of the rare earth elements, which yielded some thirty new isotopes.

From the point of view of the identification of the more abundant isotopes our knowledge is now complete. Two years ago only four elements, palladium, iridium, platinum, and gold, remained, and since then all these have been analysed by Dempster by the use of a new method employing an intense vacuum spark.

In all over 260 stable isotopes are known of which seven were discovered by observations on optical spectra, and have since been confirmed by the mass-spectograph. This large assembly shows many empirical laws, of which perhaps the more remarkable is that no odd numbered element has more than two isotopes. Even elements are not so limited. The most complex element so far observed is tin, with ten isotopes ranging in mass-number from 112 to 124. One of the most astonishing results is that for practically every natural number up to 210, a stable elementary atom is known, many are filled twice over and a few three times with "isobares", that is atoms of the same weight but different chemical properties. Schemes of tabulation of all the known species have led to the prediction of isotopes and to theories of nuclear structure to account for their occurrence.

Study of the relative abundance of isotopes in the mixture we still call, for convenience, an element, is of interest from two entirely different points of view. In the first place, since it appears to be perfectly invariable in Nature, not only in terrestrial but also in meteoric matter, there was a slight hope that a systematic measurement of abundance ratios might disclose some simpler relations bearing on the great problem of how the nuclei of atoms were evolved. The relative abundance of isotopes can be estimated by several methods, but that of the most general application is the photometry of mass-spectra. A technique of this was worked out in 1929, and a number of elements examined, but the ratios, obtained in numbers large enough for statistical treatment, showed no groupings other than would have been expected from pure chance. These measurements have a second important practical value. If we know the masses of the isotopes of an element and their relative abundance it is easy to calculate their mean weight. This, with proper corrections, can be used to check the chemical atomic weight. During the past six years nearly every atomic weight has been determined by this purely physical method, which has the great advantage of being, in general, independent of purity, and requiring an almost infinitesimal quantity of material.

Instead of the original view that the nuclei of atoms consisted of protons and electrons, it is now considered more likely that they are built of protons and neutrons. In either case the binding forces holding the particles together must represent loss of energy, that is, loss of mass. Hence it is that the atom of hydrogen has abnormally high mass, and that the accurate determinations of divergences from the whole number rule are of such profound theoretical importance. As I have stated, my second mass-spectograph was designed for this and found capable of an accuracy, in favourable cases, of 1 in 10,000. The atom of oxygen 16 was chosen as standard and the percentage divergences, expressed in

parts per 10,000 called "packing fractions", were
determined for a large number of elements. These,
when plotted against mass-number, were found to lie
roughly on a hyperbolic curve. This drops rapidly from
hydrogen, passes through a minimum of about -10
in the region of iron and nickel, and then rises gradually,
crossing the zero line in the region of mercury. Our
knowledge in this field has been notably increased by
the brilliant work of Bainbridge, who set up at Swarth-
more a powerful mass-spectograph of an original design
which made use of a velocity selector and semicircular
focusing. With this instrument he discovered new isotopes
of tellurium, rectified results on zinc and germanium, and
has made many of the most accurate comparisons of
mass so far known.

The events which led up to the discovery of the
remarkable isotope "heavy hydrogen" are of particular
interest. The first accurate comparisons of the masses,
now termed "isotopic weights" of the atoms ^1H, ^{12}C,
^{14}N with the standard ^{16}O were made with my second
mass-spectograph and published in 1927. The mass of
^1H could only be obtained indirectly through the intermediate mass ^4He, and was given as $1 \cdot 00778$. This and
the others agreed very accurately with the atomic
weights of the elements obtained by chemical means.

This satisfactory agreement was completely upset in
1929 by the startling discovery of the heavy isotopes of
oxygen 17 and 18 which, present in small quantity,
had naturally been overlooked on mass-spectra of that
element owing to the technical difficulty of ensuring the

absence of the isobaric compound lines OH and OH_2. The discovery was made by Giauque and Johnson by observations on band-spectra, which are free from this confusing disability, and the careful quantitative work of Mecke, made later, showed that, owing to the presence of these isotopes, the chemical standard of atomic weight $O = 16$ was about 2 parts in 10,000 heavier than the physical one $^{16}O = 16$. Examination of compounds of carbon and of nitrogen by the same method showed not only that these elements also contained heavy isotopes ^{13}C and ^{15}N but that their apparent abundance, by a most incredible coincidence, was just about enough to bring their mean weights into line with that of oxygen.

Birge pointed out that to satisfy my low estimate of ^1H hydrogen must also contain at least one heavy isotope. Urey took up the problem and, happily unaware of the real uncertainty in the figures concerned, with the collaboration of Brickwedde and Murphy fractionated liquid hydrogen and proved by examination of the Balmer lines that ^2H was present. Washburn showed that its heavier atoms could be concentrated by the electrolysis of water. This method was developed so rapidly and brilliantly by Lewis that, soon after its discovery, pure heavy water had been obtained in appreciable quantity. The isotope of hydrogen of mass 2 cannot be treated as a normal isotope. Its exceptional difference in mass enables it to be separated with comparative ease in a pure state. It has been given the name deuterium, symbol D, and heavy water D_2O is now obtainable in quantity at reasonable prices, one of the most surprising reagents in the history of science and certainly one which would have dismayed the founders of the C.G.S. system of units.

Now that deuterium is available the mass of ^{1}H can be measured with far greater certainty by the "doublet" method and turns out to be 1·00812. Now had that value been obtained at first it is quite possible that no one would have looked for heavy hydrogen, so it was very fortunate that the mistake was on the right side.

The only moral to be drawn from this seems to be that you should make more, more and yet more measurements. Even a bad one *may* be of service, but, fortunately, it will be essential for you to make a considerable number of good ones first, or no notice will be taken of it.

In recent years the accuracy of measurement has been steadily improving. In my third instrument which has second-order focusing, and in a still more powerful double focusing mass-spectrograph constructed by Bainbridge at Harvard, it approaches 1 in 100,000. The need for this high accuracy is in the recently discovered artificial transmutations, the nuclear chemistry of the future. The equations of this can only be founded upon accurate knowledge of the masses concerned. Armed with such knowledge the nuclear chemists, I am convinced, will be able to synthesise elements just as ordinary chemists synthesise compounds, and it may be taken as certain that in some reactions sub-atomic energy will be liberated.

There are those about us who say that such research should be stopped by law, alleging that man's destructive powers are already large enough. So, no doubt, the more elderly and ape-like of our prehistoric ancestors objected to the innovation of cooked food and pointed out the grave dangers attending the use of the newly discovered agency, fire. Personally I think there is no doubt that sub-atomic energy is available all around us, and that one day man will release and control its almost infinite power. We cannot prevent him from doing so and can only hope that he will not use it exclusively in blowing up his next door neighbour.

VI. FORTY YEARS OF ASTRONOMY

by

SIR ARTHUR EDDINGTON

Plumian Professor of Astronomy,
Cambridge

ASTRONOMY

IF we compare astronomy to-day with astronomy forty years ago, we notice a centrifugal tendency. Astronomical books in the 'nineties generally gave a full account of the sun, moon, planets, and comets, but could find little to tell us about the fixed stars. The centre of interest has now shifted from the solar system to the system of the stars, and beyond. We are likely to-day to hear more about galaxies than about planets.

This contrast may give a misleading idea of the pursuits of astronomy forty years ago. Actually the systematic observation of the stars was then absorbing the main energy of observatories, as it had been doing throughout the eighteenth and nineteenth centuries. In some of the work the stars were observed, not as celestial objects interesting in themselves, but as auxiliary to other researches; they were the reference marks, or graduations on the dial of the heavens, whose positions must be precisely calibrated to serve for measuring the movements of planets and comets. But besides routine observation of position, there were many kinds of investigation more directly concerned with the nature of the stars themselves. In 1900 the position was that, although actual results were still scanty and uncertain, the labours of many decades were approaching fruition and a rapid advance was in prospect.[1]

[1] An authoritative account of our knowledge at this time will be found in Newcomb, *The Stars. A Study of the Universe* (John Murray, 1902).

For an understanding of the system of the stars the first essential is a general knowledge of stellar distances. The distances of two of the nearest stars, α Centauri and 61 Cygni, were determined in 1839; but there had not been very much advance in our knowledge by 1900. The number of well-determined parallaxes had increased to about 20; but these were not discriminated from a large number of inferior and quite misleading determinations. Even in 1910 the generally accepted catalogue of parallaxes listed 88 stars within 10 parsecs distance of the sun, of which 40 have since been rejected as beyond the limit. We knew very little about the distances of individual stars, and—much worse—we did not know how little we knew.

Soon after 1900 a great improvement was made by determining stellar parallaxes photographically instead of with a heliometer. The pioneers of the photographic method were Schlesinger at the Allegheny Observatory and Russell and Hinks at Cambridge. When the distance of a star has been found by parallax determination, its apparent brightness can be converted into absolute (intrinsic) brightness. The number of well-determined parallaxes now available is sufficient to give a good idea of the range of absolute brightness of the stars, and of the dependence of brightness on spectral type and other characteristics. The direct method of determining distances is of very limited application, since most of the stars are much too remote to give measurable parallax. But from knowledge of the nearer stars (within, say, a hundred light-years of the sun) derived in this way we have obtained a firm basis for developing and controlling a number of indirect methods of estimating the distance and absolute brightness of more remote stars—so that, by a series of steps, distances of objects up to 300 million light-years are now ascertainable.

Another type of measurement which developed very slowly was the determination of the radial velocities of stars. In principle the rate of approach or recession of a luminous body in the line of sight can be found by measuring the Doppler shift of its spectral lines. This was first applied to the stars by Huggins in 1868; but by 1900 we still did not know the radial velocity of any star. If any good determinations then existed, they were buried among the totally erroneous determinations. The practical development of the technique of this measurement is due to Campbell at the Lick Observatory. In 1913 trustworthy radial velocities of 1400 stars were available.

One kind of datum was fairly abundant in 1900, namely proper motion, i.e. the apparent angular motion of a star across the sky. These motions are larger than is often supposed; and with modern appliances it is the exception for a star to show no detectable motion in twenty years. The fastest motion is that of a faint star of magnitude 9·7 known as Munich 15040 (or less officially as Gilpin) which covers 10·3 seconds of arc per year; it would be just possible to detect its motion between two consecutive nights. Our knowledge of proper motions up to 1900 rested mainly on Bradley's observations of the positions of some 3000 bright stars around 1755, which were compared with the modern positions. Whenever a new catalogue of proper motions was produced, it seemed to be taken for granted that the one use for it was to determine the Solar Apex, i.e. the direction of the sun's motion relative to the system of the stars. You might think that this occupation would pall after a time; but happily each new determination disagreed with the older ones, and thus gave astronomers plenty to talk about.

In a general way the effect of the sun's motion amid the stars is quite conspicuous. If we plot the proper motions of stars in a small region of the sky we notice at once a preponderance in one direction; the stars in the mean are moving in that direction relatively to the sun. This is expressed equivalently (but more modestly) by saying that the sun is moving in the opposite direction relatively to the stars. But though the general effect is plainly seen, an exact determination of the direction is difficult, since the apex is sensitive to small systematic errors in the system of the proper motions which we are still endeavouring to eliminate.

Just at the time I entered astronomy (at Greenwich in 1906) a revolutionary discovery had been made. The greatest pioneer in the study of stellar statistics was Prof. J. C. Kapteyn of Groningen. There is now a long series of "Groningen Publications" relating to these problems. The most interesting of them all is No. 6. But it is no use going to a library to consult it; for the interesting thing about No. 6 is that it was never written. Nature took an unexpected turn, and would not fit into the scheme which No. 6 was promised to elaborate. No. 5 was entitled "The distribution of cosmic velocities: Part I, Theory"; it was a study of how the motions of stars pursuing their courses at random would turn out statistically when the solar motion and effects of varying distance were allowed for. Meanwhile the observed Auwers-Bradley proper motions were being prepared for comparison, so as to determine the numerical constants in the formulae. But the theory, though it represented the unquestioned views of the time, turned out to be so wide of the mark that not even the beginnings of a comparison were possible; and the application of the formulae had to be abandoned. This was Kapteyn's great discovery of the two star streams, announced at the British Association meeting in South Africa in 1905, which revealed for the first time a kind of organization in the system of the stars and started a new era in the study of the relationships of these widely separated individuals.

At first this discovery was received with much in-credulity; but to any one who took the trouble to examine the proper motions for himself there could never be any doubt. For example, the diagram overleaf shows the statistics of distribution of the proper motions in a typical region of the sky; the radius from the origin to the curve in any direction is proportional to the number of stars whose motions are in that direction. You can see easily that there are *two* favoured directions of motion, indicated by the full arrows. The direction towards the solar antapex is shown by the broken arrow; although the stars in the mean move in this direction, they do not favour it individually. Evidently the double streaming cannot be accounted for by the parallactic effect of the sun's own motion; it is an intrinsic peculiarity in the distribution.

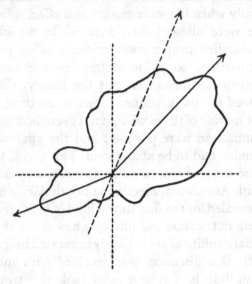

The phenomenon of two star streams has been abundantly confirmed; and it is found to prevail everywhere in the region over which our surveys of proper motions extend. Its real significance is still uncertain. It may be that in our neighbourhood two large clusters of stars have met by accident and are passing through one another. But in modern discussions star-streaming has to be considered in conjunction with another effect whose existence was demonstrated by Oort of Leyden

about 1927. The whole galaxy is found to be rotating
about a centre which lies 10,000 parsecs away in the
direction of the constellation Sagittarius. This centre is
fairly near to the vertex of the preferential motion;
though according to the best determinations there is a
difference of about 10°, which seems to be real. It
suggests itself that Kapteyn's phenomenon is primarily
a tendency for the stars to move radially (towards or
away from the centre of the galaxy) rather than trans-
versely, and that it is a general feature of the dynamics
of the system.

In accordance with the law of gravitation it is
necessary that the rate of revolution of the galaxy should
decrease outwards; just as in the solar system the outer
planets revolve more slowly than the inner planets.
Our survey of proper motions and radial velocities
extends over what is after all only a small part of the
whole galaxy; consequently in actual observation we
are concerned only with the *differential* rotation in this
small region. Since the outer part of a region travels
more slowly than the part nearest the galactic centre,
the region must become distorted. Considering a square
with ourselves at the centre, the square will become
sheered into a diamond. The stars seen in the two
opposite directions corresponding to the lengthening
diagonal will therefore (on the average) be moving away
from us; those 90° away, corresponding to the con-
tracting diagonal, will be coming towards us. The
sheering effect, by which motions of approach and
motions of recession follow at intervals of 90° of galactic

longitude—or, as it is briefly described, the double-period term in the average radial velocity—is shown by all classes of distant stars. The results of the analysis, giving the direction of the centre and the magnitude of the differential rotation, are very accordant. The rotation indeed provides a means of estimating the average distances of various classes of stars; since the amplitude of the double-period term, which we obtain by analysis of the observed radial velocities, is proportional to the distance. This illustrates one of the indirect methods by which our knowledge of celestial distances has been extended beyond the limits within which parallax measurement is possible.

We are a long way from the centre of the stellar system—perhaps about half-way out towards its confines. In our neighbourhood the orbital speed is about 250 kilometres per second, and the time required for a complete revolution is of the order 250 million years. It is interesting to reflect that we have made five or six complete revolutions round the system within geological times. It is also possible to infer the mass of the system, which controls this orbital motion. It must be about 150,000,000,000 times that of the sun; so that presumably the system contains something like that number of stars. This is about ten times the number estimated by extrapolation of actual counts of stars; but it is not incredible, because vast numbers of stars may be hidden by the clouds of obscuring matter which are observed in various parts of the system and especially in directions towards its centre.

To return to the state of our knowledge in 1900—it was realized that the stars form a limited system strongly flattened in the galactic plane like a bun or lens. Newcomb, who was one of the leading authorities, concluded that the boundary of the stellar universe was nowhere distant more than 3000 light-years and might possibly be much less. The diameter now assigned to it is at least 20 times greater. But that is by no means the whole magnification that our conception has undergone. This huge system of a hundred thousand million stars is only the beginning; it is one island galaxy among many. Out beyond it there are other islands, recognized by us as spiral nebulae. From sample counts it is computed that 10,000,000 of these galaxies, coequal with our own island system, are detectable with our present telescopes; and doubtless there are far more outside their range. In forty years the "boundary" of the material universe has been pushed back from 3000 light-years to beyond 500,000,000 light-years, which is the present limit of telescopic survey.

The status assigned to the spiral nebulae has undergone remarkable vicissitudes. The hypothesis that the nebulae are island galaxies was first put forward by Sir William Herschel, and for a time was widely accepted. When the spectroscope was introduced, Huggins found that many of the most conspicuous nebulae show the bright-line spectrum of a gas and are clearly not galaxies of stars. This discredited the island theory; and all the nebulae were brought inside our galactic system again. Later it became clear that the spiral or "white" nebulae are a different class of objects from the nebulae with gaseous spectrum. About 1910 the island theory was revived as regards the spiral nebulae. It was advocated especially by Sir David Gill and an international group of astronomers associated with him; but Van Maanen, Jeans and others took the opposite view, and the question was strongly contested until 1924, when Hubble produced conclusive evidence that the spiral nebulae are extragalactic. The gaseous nebulae on the other hand are intragalactic.

In some of the nearer spiral nebulae it is possible to recognize individual stars. Hubble found among them some Cepheid variables, and determined the periods of their light-variation. It is known from the study of Cepheid variables in our own galactic system that the period is a trustworthy indicator of the intrinsic brightness. Thus the absolute light-power of these stars could be assigned; and, by comparing it with their apparent brightness, the distances of the Cepheids, and of the nebulae containing them, were deduced. These distances showed them to be far outside our own galaxy.

The spiral nebulae are found to be moving away from us systematically with speeds approximately proportional to their distances. This indicates that the whole system of the galaxies is dispersing. At the present rate of expansion its dimensions will become doubled in 1500 million years—a period comparable with the age of the earth's crust. The rapidity of this change of the large-scale structure of the universe has profoundly modified our ideas of the time-scale of evolution in astronomy. Upper limits to the ages of the stars had been calculated on the extreme assumption that the whole of the energy of constitution of the matter contained in them is available for radiation; for example, the upper limit for the sun is five billion (5.10^{12}) years. Formerly it was widely believed that the actual ages correspond to these upper limits; but such values are now altogether incongruous. We cannot well assign to the stars a past duration more than a small multiple of the "time of relaxation" of the system of galaxies. For this and other reasons the age is now put not higher than 10^{10} years.

I dare not enter further into the discussion of this expansion of the universe, lest I be tempted to spend the rest of the lecture over it. But in view of the false impressions which exist, it is desirable to emphasize that it is a straightforward development of astrophysical research in which observation and theory very happily agree. The cause of the expansion is a force known as "cosmical repulsion" which is foretold by the equations of relativity theory. At very great distances Newtonian attraction is modified by this supplementary force, and in the system of the galaxies the resultant effect is a repulsion which scatters the galaxies apart. It would be contrary to the scientific spirit to represent this, or any other result described in this lecture, as safe from revision through the progress of knowledge. But it is not one of the more precarious advances; and there is no excuse for those who treat cosmogony as a field for unguided speculation of a type which would not be tolerated in any other branch of physical science. I think that a joint paper by Einstein and de Sitter a few years ago must be held partly responsible for this wave of speculation. It was a piece of mathematics, innocuous in itself, but put in such a way as to give the impression that these distinguished authorities had become sceptical about the cosmical constant. Einstein came to stay with me shortly afterwards, and I took him to task about it. He replied: "I did not think the paper very important myself, but de Sitter was keen on it." Just after Einstein had gone, de Sitter wrote to me announcing a visit. He added: "You will have seen the paper by Einstein and myself. I do not myself consider the result of much importance, but Einstein seemed to think that it was."

To return to historical order, after the discovery of the star streams the next big sensation in stellar astronomy was the Giant and Dwarf theory put forward by Hertzsprung and Russell, which came into prominence about 1913. In 1900 we were supposed to understand thoroughly the course of stellar evolution—the stars passed through the sequence of spectral types B, A, F, G, K, M in order, ending up as dark stars. The description of spectral types as "early" or "late", which is still commonly used by force of habit, is a reminder that at one time the above sequence was accepted with the utmost confidence. But whereas in most branches our knowledge has greatly advanced, our knowledge of stellar evolution seems to have diminished, until now it is represented approximately by the symbol o.

Apart from any evolutionary interpretation, the order

B, A, F, G, K, M is that of diminishing temperature; the supposed passage of a star down this sequence therefore signified that its surface continually cooled. Hertzsprung and Russell discovered that each of the "later" types comprises two distinct classes of stars, widely different in absolute brightness and presumably also in physical condition. Within any one spectral type the surface luminosity must be approximately the same; thus the difference of brightness can only be attributed to difference of size. For example, the M stars are divided into a group of huge diffuse "giants" and a group of small concentrated "dwarfs"—with no intermediate M stars. As we go up the spectral sequence the giant and dwarf groups converge, and they coalesce in types A and B. To put the stars in order of increasing density, we must begin with the M giants, go up the sequence M, K, G, F along the giant branch to A or B, and then come down the dwarf branch to M again. Hertzsprung and Russell took this to be the order of evolution, the stars being supposed to begin as condensations in nebulae and to contract continually as they radiated away their heat.

It had been shown by Homer Lane in 1870 that when a gaseous star contracts its internal temperature must rise; this seemed to account for the ascending part of the evolutionary sequence which corresponds to diffuse giant stars. At the turning point in type B or A the density is beginning to approach that of water. It was naturally thought that the gas theory then ceased to apply; the second half of the evolutionary sequence accordingly followed, in which the star descended the dwarf branch cooling like a liquid or solid mass. We notice that the keystone of the Hertzsprung-Russell theory of evolution was that the change over from ascending to descending temperature occurs when the star becomes too concentrated to behave as a perfect gas. In view of the subsequent collapse of this theory we may call attention to a weak point in the argument. It confused the *internal* temperature (treated in Lane's theory) with the *surface* temperature (indicated by the spectral type); these do not necessarily increase and decrease together.

The distinction of giants and dwarfs was a very important advance. In modern classification there is a slight modification. We recognize three groups: firstly, the Main Series, which is the old dwarf branch prolonged to include types A and B and a still hotter type O; secondly, the Giants, which are not very numerous in space, but owing to their great brilliance furnish the majority of the naked-eye stars; thirdly, the White Dwarfs, which fall on the other side of the main series, and have transcendently high density. But the evolutionary part of the theory has been abandoned altogether. It was my lot to bring about its fall in 1924 by showing that the dwarf stars, notwithstanding that their density is as high as that of terrestrial liquids and solids, are still in the condition of a perfect gas and have the compressibility of a gas. That contradicts what I have called above the keystone of the theory. Another upsetting result was that the absolute brightness of stars (other than white dwarfs) is determined almost wholly by their masses and is little affected by their density; in particular, a dwarf has much smaller mass than a giant of the same type, so that we cannot take them as representing later and earlier stages of the same star—unless indeed a star loses mass as it grows older. For a time it seemed possible to save the Hertzsprung-Russell sequence of evolution by postulating that a star during its lifetime radiates away a large proportion of its mass. But this suggestion, which involves the hypothesis that electrons and protons can annihilate one another, has been undermined by the recent discoveries in atomic physics; and, moreover, the shortened time-scale already mentioned gives insufficient time for the required decrease of mass by radiation.

Certain distinctive characteristics in the spectra of giant and dwarf stars have been found, so that they can now be distinguished easily without examining their absolute brightness. By an elaboration of the same principle Adams and Kohlschütter developed in 1914–16 a spectroscopic method of determining absolute brightness. As already stated the stars can be arranged in sequence in order of surface temperature, certain features of their spectra being used as criteria. Other less obvious features give a classification which runs athwart the temperature classification. It is found that this transverse classification is governed by the absolute brightness. This gives an empirical method of determining the absolute brightness, stars whose absolute brightness is already known being used to calibrate the scale. By comparison with the apparent brightness we can deduce the distance or parallax. The results of this method are therefore called "spectroscopic parallaxes".

In 1920 the angular diameter of a star was measured for the first time by Pease and Anderson with an interferometer constructed by Michelson. The stars are so distant that no disc can be seen even in the largest telescope; the image is always indistinguishable from that of a point. Actually it is not difficult to compute their angular diameters from their apparent brightness and the approximately known effective temperatures of the different spectral types; but a check on the theory is always desirable, particularly as in this case it led to what appeared to be outrageously improbable results for the companion of Sirius and other white dwarfs. Probably no star has a disc greater than $\frac{1}{20}$th of a second of arc; the largest should be those of the first magnitude stars of type M, namely Betelgeuse and Antares. Using apertures separable up to 20 feet, after the pattern of a range-finder, Pease and Anderson obtained diffraction effects for Betelgeuse from which a diameter of 0″·045 was deduced. This gives a linear diameter of about 300 million miles; that is to say, Betelgeuse is large enough to contain the orbit of Mars. To extend the method to somewhat smaller stars a 50-foot interferometer was built; but it was not brought into successful use until 1936.

I became interested in the theory of the interior of a star in 1915. The hypothesis that the variable stars of a particular class known as Cepheids are pulsating stars, first proposed by Plummer, had recently received strong support from the researches of Shapley; and I wanted to investigate the thermodynamic problems connected with the maintenance of a pulsation. But it was not until some years later that I was in a position to begin the investigation of pulsations; for it soon became clear that the existing theory of a static star needed to be modernized to correspond with advances which had been made in the physics of radiation.

You must understand that by "modernized" I do not mean that I intended anything so revolutionary as the introduction of quantum theory. I suppose I believed in quantum theory after a fashion—in so far as anyone could believe the tissue of inconsistencies that it then was. But it was not the sort of thing for a matter-of-fact astronomer to have dealings with. You may find it difficult to recall the outlook of those benighted days. Let us go back to 1912. At that time quantum theory was a German invention which had scarcely penetrated to England at all. There were rumours that Jeans had gone to a Conference on the continent and been converted; Lindemann, I believe, was expert on it; I cannot think of anyone else. Soon afterwards Bohr's theory of the atom appeared, and quantum theory reached the stage at which one says: "There seems to be something in it, and I really must read it up when I get time." By 1915 everyone was reading Jeans' "Report on Quantum Theory", published by the Physical Society; and upholders of the "pint-pot" type of explanation were being driven into the last ditch.

So the theory was very much under discussion. But as you must have realized, I am always very conservative in my outlook and do not accept anything until it has become quite obvious that it is true; and I did not want the work I was doing on the structure of the stars to be mixed up with half-baked theories.

However, a point soon arose where quantum theory could not be evaded. Following all earlier writers I had taken the ultimate independent particles in stellar material to be atoms, and had adopted appropriate masses for them. Newall first suggested to me that it was more likely that there would be complete ionization, setting free all the satellite electrons; if so the average mass per particle would be much lower. The quantum experts Jeans and Lindemann were of the same opinion. It seems to have been a matter of opinion only, for there was no recognized quantum theory of ionization at that time. I found the suggestion attractive, because it made my numerical results nearly independent of the unknown chemical composition of the stars, the average molecular weight of fully ionized matter being nearly the same for all elements except hydrogen. Calculations were therefore made both on the assumption of high ionization and of low ionization, it being left to observational comparison, or to the future development of theory, to decide between them.

The present ionization formula, as given by quantum theory, was first applied to the stars by Eggert in 1919; and it confirmed the hypothesis of high ionization which was already provisionally assumed. In 1921 Saha studied in the same way the state of ionization of the outer layers of a star, and started the modern theory of the interpretation of stellar spectra. This has opened out a very wide field of practical and theoretical investigation. It is in fact the parent of most of the problems of modern astrophysics.

One outcome of the theory of the interior of a star was the realization that matter in dwarf stars, though comparable in density with iron or water, still behaves as a perfect gas and is therefore far from the limit of compressibility. There is no obstacle to densities hundreds or thousands of times greater. It suggested itself that this was the explanation of a difficulty presented by the white dwarf stars. The companion of Sirius, for example, gives only $\frac{1}{360}$th of the light of the sun, although judging by the spectral type its surface should radiate more strongly. The star must therefore be much smaller than the sun, the computed diameter being no more than 25,000 miles. The resulting density, about 60,000 times the density of water, had appeared to indicate a fallacy in the deduction, and had cast suspicion on the reliability of spectral type as a guide to temperature. But on the new theory the high density was not incredible, and the calculation might be accepted literally.

To decide the question Adams investigated the Einstein shift of the spectral lines in the companion of Sirius. The shift, which is proportional to the gravitational potential at the surface of the star and therefore varies as mass divided by radius, would be very large if the small radius were correct. In 1925 the large shift was observed, and the high density was confirmed. Even this is not the highest stellar density; although only rough estimates can be given for the other known white dwarfs, it appears that in some of them the density must be at least 100 tons to the cubic inch.

Up to 1924 no one seems to have given serious thought to the possibility of superdense matter. By a strange coincidence, just when astronomers were discovering its existence in the white dwarf stars, the thoughts of pure physicists were independently turned in the same direction. Wave-mechanics was found to introduce deviations from the classical statistical theory of particles, which in general could only be appreciable at extremely high density. To those seeking an application of the new theories the white dwarfs were a godsend. The new (Fermi-Dirac) statistics were first applied to them by R. H. Fowler, who thereby cleared up a serious difficulty as to the energy-content of the white dwarfs.

In the later part of the period here surveyed the greatest advance perhaps has been in the study of gaseous nebulae. These are extremely rarefied objects with density generally of the order 10^{-20} gm. per cu. cm. Some of them (planetary nebulae) surround one star only; others (irregular nebulae) are much more extensive and enclose many stars. In all cases one or more stars of very high temperature must be enclosed, because the nebular light is due to stimulation by extreme ultra-violet radiation from the stars. The light is not a simple reflection but a fluorescence effect.

For many years the principal puzzle was the nature

of the bright spectral lines responsible for the light of the nebulae, since the most conspicuous lines could not be traced to any known element. A new era began in 1927 when Bowen found that a number of the unknown lines were "forbidden" lines of ionized oxygen and nitrogen. In fact the material which had puzzled us was just air. These forbidden lines have not been observed in the laboratory; but their theoretical wave-lengths can be calculated from those of the ordinary lines in the spectrum of the same element and identified with the observed nebular wave-lengths. They represent downward transitions from a metastable state, that is to say a state in which the atom will usually remain a long time (from a second to several hours) before it falls to a lower state with emission of radiation. Consequently to emit forbidden lines the atom needs to be left undisturbed for long periods—a condition impossible to secure in the laboratory but satisfied in the nebulae, where the sparsely strewn atoms and electrons travel for hours without colliding, and the radiation is so dilute that encounters with photons are equally rare.

Following up this result Zanstra produced a very fascinating theory which traces in detail the transformations of radiation in the nebula, enables the effective temperature of the stimulating star to be calculated, and accounts for the principal facts observed. Unfortunately it is too lengthy and technical to be described here.

I need scarcely say that in this brief survey of forty years' progress in astronomy much must be omitted; I must not, however, neglect the solar system entirely.

Its recognized membership has increased by four more satellites of Jupiter, one of Saturn (discovered in 1898), 940 minor planets (bringing the total to 1380), and one major planet Pluto. The trans-Neptunian planet Pluto was discovered by Tombaugh in 1930. It has a period of 248 years as compared with Neptune's 165 years, and a mean distance from the sun 3600 million miles compared with Neptune's 2800 million miles; but its orbit is so eccentric that at perihelion it comes slightly nearer to the sun than Neptune. Estimates of its diameter and mass are still only hypothetical; but it seems certainly smaller than the earth, and may perhaps be about the size of Mercury. It was found in the course of a search made at the Lowell Observatory for a trans-Neptunian planet predicted by the late Prof. Lowell from a study of the perturbations of Uranus. It turned up close to the predicted spot. But it has since been shown that it could not have been responsible for the perturbations, and the agreement must be set down as pure coincidence.

Perhaps the most generally interesting advance in this part of astronomy has been the spectroscopic study of the atmospheres of planets. Our familiar atmosphere of oxygen and nitrogen, with a certain amount of water-vapour, is not by any means a standard equipment. The atmosphere of Mars, though rather scanty, is not dissimilar to that of the earth; but all the other planets show significant differences. Mercury (like the moon) shows no trace of an atmosphere. Venus is covered by thick cloud and only the upper atmosphere above the cloud can be examined; its spectrum shows carbon dioxide, but no oxygen. It is thought that this signifies that there is no vegetation of the terrestrial type on Venus; since on the earth vegetation performs the function of absorbing carbon dioxide from the atmosphere and returning the oxygen. This is in keeping with another line of speculation which had suggested that the surface of Venus is entirely covered by ocean. On Jupiter and Saturn we find atmospheres of an entirely different constitution, namely a mixture of methane and ammonia—which seems an extraordinarily unpleasant combination! The proportion of ammonia is less on Saturn than on Jupiter; and on the still more remote planets Uranus and Neptune only methane has been detected. This progressive decrease of ammonia may be attributed to the increasing cold which removes it from the atmosphere by liquefaction.

The presence of hydrides (CH_4 and NH_3) in the atmospheres of the four large planets suggests an excess of hydrogen in their composition. In the last ten years we have come to realize that hydrogen is an extremely abundant element. Studies of the interior of a star, of the surface layers of stars, and of the gaseous nebulae, have all independently led to this conclusion; and one of the leading factors in the progress of the last ten years has been the revision of older views which had not appreciated the important part played by this element. It can generally be taken that hydrogen is nearly as abundant by mass as all the other elements put together, and far more abundant if we reckon by number of atoms. We therefore assume that when the planets were formed from the sun, the material originally contained abundant hydrogen. That it is not so important a constituent of the earth is explained by the theory of escape; the earth's gravitational field was not strong enough to prevent the light and fast-moving atoms of hydrogen from leaking away. But the massive outer planets have a stronger field and could retain their hydrogen. This gives a satisfactory explanation of the different constitution of their atmospheres; and a detailed chemical study of the problems raised throws much light on the physical condition of these planets. I will only add here that the conditions are intensely cold, and all the water must lie ice-bound on the planet's surface—concealed from us and from the sun's rays by the deep opaque atmosphere.

I will devote the few minutes which remain to the most exciting event I recall in my own connection with astronomy, namely the verification of Einstein's prediction of the deflection of light at the eclipse of 1919. The circumstances were unusual. Plans were begun in 1918 during the war, and it was doubtful until the eleventh hour whether there would be any possibility of the expeditions starting. But it was very important not to miss the 1919 eclipse, because it was in an exceptionally good star-field; none of the subsequent expeditions have had this advantage. Two expeditions were organized at Greenwich by Sir Frank Dyson, the late Astronomer Royal, the one going to Sobral in Brazil and the other to the isle of Principe in West Africa. It was impossible to get any work done by instrument-makers until after the armistice; and, as the expeditions had to sail in February, there was a tremendous rush of preparation. The Brazil party had perfect weather for the eclipse; through incidental circumstances, their observations could not be reduced until some months later, but in the end they provided the most conclusive confirmation. I was at Principe. There the eclipse day came with rain and cloud-covered sky, which almost took away all hope. Near totality the sun began to show dimly; and we carried through the programme, hoping that the conditions might not be so bad as they seemed. The cloud must have thinned before the end of totality, because amid many failures we obtained two plates showing the desired star-images. These were compared with plates already taken of the same star-field at a time when the sun was elsewhere, so that the difference indicated the apparent displacement of the stars due to the bending of the light-rays in passing near the sun.

As the problem then presented itself to us, there were three possibilities. There might be no deflection at all; that is to say, light might not be subject to gravitation. There might be a "half-deflection", signifying that light was subject to gravitation, as Newton had suggested, and obeyed the simple Newtonian law. Or there might be a "full deflection", confirming Einstein's instead of Newton's law. I remember Dyson explaining all this to my companion Cottingham, who gathered the main idea that the bigger the result, the more exciting it would be. "What will it mean if we get double the deflection?" "Then", said Dyson, "Eddington will go mad, and you will have to come home alone."

Arrangements had been made to measure the plates on the spot, not entirely from impatience, but as a precaution against mishap on the way home; so one of the successful plates was examined immediately. The quantity to be looked for was large as astronomical measures go, so that one plate would virtually decide the question, though, of course, confirmation from others would be sought. Three days after the eclipse, as the last lines of the calculation were reached, I knew that Einstein's theory had stood the test and the new outlook of scientific thought must prevail. Cottingham did not have to go home alone.

I have told you, as best I can, something of what we have learnt in the last forty years. I will end by expressing a hope—not unmingled with doubt—that not too much of what I have been saying will be upset in the next forty years.

VII. FORTY YEARS OF PHYSIO-LOGY AND PATHOLOGY

(BEAUMONT, PAVLOV AND AFTERWARDS)

by

JOHN A. RYLE
Regius Professor of Physic, Cambridge

PHYSIOLOGY AND PATHOLOGY
(BEAUMONT, PAVLOV AND AFTERWARDS)

In the course of this lecture I shall seek to recall and to illustrate some of the more fruitful interrelationships of the medical sciences. To have confined my period for review within the strict limits prescribed by the general scheme and title of this course would have required the omission of some important sequences, and I must therefore ask to be allowed a longer retrospect. Nevertheless, the fragments of medical and physiological history which I have pieced together are, for the most part, drawn from chapters chronologically recent or frankly modern. These serve to illustrate in a variety of ways what may be described as the importance of continuity.

For reasons which will become apparent I have entitled the lecture "Beaumont, Pavlov and Afterwards". Pavlov properly opens the period, but his work could not be discussed without reference to the work of Beaumont sixty years earlier. Neither Beaumont nor Pavlov, for all the completeness of their observations and careful preparation of the ground, could have foreseen some remarkable discoveries, within their own field of study, which have derived from recent clinical investigation.

History, whether political, national or scientific, is largely woven around personalities and movements due to personalities. Scientific history is particularly concerned with movements of ideas, discoveries of fact, and the integration of new knowledge. Political and national histories, on the other hand, are both concerned with and swayed by human passions as well as human thought, and are for these and other reasons more complex and less consecutive.

If we select two of the greatest figures in medical history—Hippocrates and Harvey—we find that, apart altogether from their numerous special contributions to knowledge, both were instrumental in promoting ideas and indicating methods without which biological science, as we know it, could scarcely have been born and certainly could not have prospered. Hippocrates taught the importance of accurate observation of natural phenomena, and was the founder of the naturalistic approach to the study of man in health and disease. Harvey was the founder of the experimental method, a physician by training, but the first great systematic physiologist.

Advancement in medical learning has depended upon the simultaneous but not always parallel development of the observational and experimental schools. Rapid and steady improvements in technique, refinements in the actual tools of experiment, the interactions of the more precise sciences, and the inevitable and necessary specialistic trend have between them provided a continuous impulse for the newer method, but in the process have brought about a certain cleavage between the two schools which can be of advantage to neither.

A study even of so small a fragment of medical and

physiological progress as may be comprehended within the two words "gastric history", reveals at once the close and intimate interdependence of the observational and experimental schools, reminds us of elementary but important steps and stages which, in the abundance of our modern equipment, we are apt to forget, and adds character and even romance to a chapter of knowledge which the necessary discipline of science might otherwise render needlessly austere, or present-day familiarity too commonplace. As we move from personality to personality, from observational to experimental episode and back again to observation; as we trace not only the main theme of gastric physiology but also the clues which disclose the connection between the functions of the stomach and many other essential bodily processes we shall, I think, of necessity appreciate what I have just referred to as the importance of continuity. We shall behold a continuity of ideas not only in time, but influencing also the width and substance of research and repeatedly directing inquiry to subjects apparently detached from one another. We shall note in particular the continuity which exists, but which must still be made more intimate, between physiology, pathology and therapeutics.

Of the more famous students of gastric physiology two stand pre-eminent, and I shall naturally concern myself much with them and their work. I refer to William Beaumont (1785–1853) the American army surgeon, and Ivan Pavlov (1849–1936), the great Russian physiologist, whose death we mourned a short time ago.

Gastric physiology may almost be said to have been born with Beaumont's observations. Before his time the functions of the stomach were very imperfectly understood. William Hunter (1718–83), discussing current opinions, summarized the situation in his day as follows: "Some physiologists will have it that the stomach is a mill, others that it is a fermenting vat, others again that it is a stewpan; but, in my view of the matter, it is neither a mill, a fermenting vat nor a stewpan but a stomach, gentlemen, a stomach." I shall hope to remind you that the stomach is indeed a very individual organ, but with a very wide and complex range of functions and associations.

De Réaumur (1683–1757), who invented a thermometer and introduced the scale of temperatures which still bears his name, had a pet buzzard and obtained from it samples of gastric juice by persuading it to swallow small perforated tubes containing fragments of sponge. He described the taste and acid reaction of the juice and suggested experiments to test its digestive power. We should remember also that Spallanzani (1729–99) performed experiments on himself, swallowing small linen bags containing food and recovering them on evacuation and estimating the weight-loss of their contents through digestion; he also conducted experiments on animals, and established that gastric juice prevented putrefaction. Prout (1785–1850), the first English physiological chemist, showed that the acid of gastric juice was hydrochloric acid—a very important discovery. But it remained for Beaumont to prove in man and to correlate anew these piecemeal observations, adding at the same time much new knowledge and presenting a detailed and comprehensive picture of the whole cycle of gastric digestion.

This he did by seizing the opportunity provided by a serious accident with a shot-gun, which occurred on 6 January 1822 at Fort Mackenzie, a frontier outpost and trading station at the junction of Lakes Michigan and Huron. The victim of this accident, Alexis St Martin, a French Canadian, received the charge at a distance of 3 ft. under the left breast. A large part of his side was blown away and several ribs were fractured. Portions of the lung and stomach protruded in the wound; the diaphragm was torn and the stomach opened. Beaumont attended to him with infinite care for many months. He was left with a gastric fistula, but otherwise well; he was, however, destitute, without occupation or pension, and at first from physicianly compassion and to save him from being sent away to his birthplace in Canada, since the local authorities would not care for him, Beaumont adopted him and kept him in his own house. Six months after the accident a portion of the gastric mucous membrane prolapsed and, by another happy chance, formed an excellent valve which prevented the spontaneous escape of gastric contents. Ultimately in 1825 Beaumont drew up an agreement with Alexis to keep him in his employ for the purpose of his physiological experiments. These experiments continued for eight years with few interruptions, Alexis receiving his board and 150 dollars a year. When the agreement terminated, Beaumont made repeated attempts to get Alexis to come back to him for further experiments but without success.

Without a close perusal of Beaumont's work, which he entitled *Experiments and Observations on the Gastric Juice and the Physiology of Digestion*, it is impossible to obtain a true conception of his patience, his thoroughness, his resource and of the eager, accurate, scientific spirit by which all his investigations were actuated. The various sections of his little book deal with Aliment; Hunger and Thirst; Satisfaction and Satiety; Mastication, Salivation and Deglutition; Digestion by the Gastric Juice (in the stomach and *in vitro*); the Appearance of the Villous Coat (in healthy states and during indisposition or fever and after drinking bouts); Chylification and the Uses of the Bile and Pancreatic Juice. The motor, secretory and sensory functions of the stomach all received close attention. Hundreds of foodstuffs were tested with regard to their rapidity of digestion in the stomach, and also with juice removed from the stomach and placed in a vessel over a water-bath. The temperature of the stomach before and after exercise was recorded, and with the long thermometer which he used he was able to note how the pyloric portion of the stomach grasped and drew the instrument towards the pylorus, and later thrust it back again, and how, also, a distinct rotary movement was simultaneously imparted to it. He described and discussed the action of the longitudinal and transverse muscular fibres. He established that hydrochloric acid was an active principle of human gastric juice, as it had been shown to be of the juice of other animals. He also forecast the presence of some other agent to which Schwann in 1835 gave the name of pepsin. The antiseptic (or anti-putrefactive) as well as the digestive effects of gastric juice were also demonstrated.

For what must we particularly commend Beaumont beyond his experimental zeal and his patient quest for new truths? We cannot but be impressed by the promptness and sureness with which he grasped the opportunity presented to him by a rare accident. How many men would have had the requisite vision and ingenuity and persistence in the same circumstances? What a fortunate chance it was that the right man and the right opportunity were contemporary; what a fortunate chance that with Beaumont's practical skill to help him, Alexis should recover from so grave a wound and not die from shock, haemorrhage or peritonitis, or any of the possible sequels of such an injury, and still more that he should survive to enjoy full health. Without this full health the experiments could not have been regarded as "physiological".

It was genuinely a rare accident. During the whole period of the war and among a very large number of abdominal and thoracic wounds I cannot remember seeing any comparable case. Modern missiles, no doubt, have rendered appropriate wounds more certainly lethal; the long intervals which necessarily elapsed between the wound and the arrival of the soldier at a clearing-station added to their fatality; modern surgical technique no doubt repaired injuries which might have resulted in fistulae. Even so such a valvular fistula as Alexis developed clearly requires a very special sequence of chances. Without these chances and this man our knowledge of gastric physiology might have been seriously retarded and lacked completion even to this day. In Beaumont's action we have already a lesson from history, for it is certain that other accidents and diseases are still waiting to be seized upon and studied for the benefit of physiology as well as practical medicine. We are entitled to ask ourselves whether Pavlov (who naturally makes reference to Beaumont's observations), improving upon the methods of Heidenhain, would have evolved his wonderful animal experiments without the inspiration which the young army surgeon had provided.

For a truly delightful account of Beaumont and his work I would refer you to Osler's "A Backwoods Physiologist", included in the *Alabama Student*—the best, perhaps, of the many fine biographical essays included in that book.

In Pavlov we encounter another type of genius. Without the accident it is more than likely that Beaumont would have made no lasting contribution to science. Pavlov, as determined a searcher for the truth as Beaumont became, would, one is tempted to infer, have become what he became, a great experimenter, in almost any circumstance. He added to the qualities which Beaumont developed experimental ingenuity and operative skill—he was ambidextrous—and an ability in the planning of experiments which have remained unsurpassed.

Those who are familiar with his later as well as his earlier work will appreciate what a master mind he possessed. He knew from the earliest days that his surgery must be precise and aseptic, that his operations for the production of a gastric pouch (the functions of which could be studied even more minutely than the stomach of Alexis and which allowed him to obtain pure juice unmixed with food) must be technically perfect, that he must leave circulation and nerve supply intact, and maintain his animals in a state of unimpaired general health. Indeed the health and happiness of his dogs was always a matter for very particular concern. Their psychology too must needs be attentively watched and understood. His beautifully controlled work on conditioned reflexes has opened a new chapter in the objective study of psychological processes. It grew in direct sequence from the work on the digestive glands.

Pavlov's name has been particularly connected with studies of secretion directly induced by food and of so-called psychic secretion. The Russian edition of *The Work of the Digestive Glands* was published in 1897. Like Beaumont he noticed the effects and studied the digestion of a great variety of foodstuffs, but with a much greater precision and with minute measurements of the amount and rate of secretion and the rate of digestion. Like Beaumont, however, he did not confine himself to physiology, but made experiments in connection with gastric pathology and therapeutics which have had their influence on practical medicine. He was his dogs' physician as well as their surgeon and a master of technique. His breadth of view compelled this. The continuity between physiology, pathology and therapeutics appealed to him more than it has done to the majority of physiologists since his day, and is repeatedly emphasized in his writings. He inspired a large band of loyal disciples, as is evident from the extensive bibliography of the various subjects which he and they have illuminated.

In common with many of his countrymen Pavlov suffered greatly at first from the revolution, but in his latter years the Soviet authorities were fully alive to the great fame which he had brought to Russia and gave him the support and privileges which he deserved. I met him once only and had the honour of helping him into his coat; his fine rugged old features will not readily be effaced from my memory.

What contributions since Pavlov's gastric experiments have helped to advance our understanding of gastric physiology and pathology? The advent of radiology and its application by Cannon in 1897 to animal experiments by the study of gastric movements and gastric form with the opaque meal, and the later application of the method to human physiological studies by Hurst and many others, have provided important new chapters. The shape, the tone, the position, the peristaltic behaviour and emptying rate of the stomach have been closely studied. The variability of the organ in men and women of varying physique has been established by numerous observations both in healthy and sick folk. The old fixed ideas of the anatomy books have had to be abandoned. Three main types of stomach, the so-called hypertonic (or short) stomach, the orthotonic, and the hypotonic (or long) organ are recognized. Cannon observed the effects of emotion on the stomach of the cat, and Barclay, who worked at Cambridge, has made some observations on emotional reactions in man. Cannon, a physiologist, must be acclaimed as the discoverer of a method which has done more, perhaps, for clinical gastro-enterology than Laennec's discovery of the stethoscope did for the study of heart and lung disease. With the opaque meal, accurate diagnosis of gastric and duodenal ulcers, and pouches, pyloric spasm and stenosis, and of new growths are now possible in a very high percentage of cases. The clinical radiology of the alimentary tract has acquired a vast literature.

The study of form and movement was preceded by studies of secretion. John Hunter, before the days of rubber tubing, had suggested the use of an eel's skin introduced with a probang as a tube for gastric lavage. Although gastric lavage with tubes had been employed before, Leube in 1871 was the first to use the stomach-tube as a means of obtaining gastric contents for analysis. Until 1914, when Rehfuss introduced the more valuable and informative method of fractional gastric analysis, Ewald's one-hour test-meal, with estimations of gastric acidity following a meal of tea and toast, was generally employed in clinical work.

Shortly after the war the Guy's school, repeating and

extending the work of Rehfuss and his collaborators in America, conducted extensive observations with the fractional test-meal in a large series of healthy medical students with a view to establishing the limits of gastric acidity and emptying time, and the variability of the response to a standard meal of gruel. Just as the stomach was shown to vary in shape and position in health, so too a wide variability in secretory activity was demonstrated, and among healthy men examples of extreme "hyperchlorhydria" and "achlorhydria" were found.

The limits of physiological variability are always worthy of study. It is possible, but not proven, that these variations may throw some light on the question of innate predisposition to certain types of disease. The high hypertonic stomach with a high acidity is a very frequent association with duodenal ulcer; achlorhydria is an almost constant precursor of pernicious anaemia. Injections of histamine, which stimulates the secretion of gastric juice, are now frequently used in conjunction with the gastric tube. They help to distinguish a false from a true achlorhydria.

In the last twenty years the output of literature on gastric secretion has been enormous in all civilised countries. Pavlov's conclusions have found no refutation, but much has been learned about the behaviour of the human stomach in health and disease. The extension of Pavlov's correlative studies on salivary secretion, already referred to, in order to demonstrate the conditioned reflex, has instructed us in regard to another continuity, familiar to physicians in a general way—namely that which exists between psychological and physical function, between the activities of the special senses and the cerebral cortex and those of the viscera and the secretory glands. The interrelated disturbances of mind and stomach provide constant problems in practice, and await a more scientific elucidation.

The newest method of clinical study is gastroscopy. A flexible gastroscope has made it possible to examine visually wide areas of the stomach wall, thereby enhancing our appreciation of the minor changes affecting the gastric mucosa and our general knowledge of living pathology, and adding a diagnostic weapon for use in special cases where clinical analysis, radiology and the test-meal do not give us the information we require. Certain appearances with which the gastroscopist is now becoming familiar were clearly described by Beaumont a hundred years ago. Our knowledge of what Lord Moynihan called "the pathology of the living" has also been appreciably advanced by the gastric surgeons.

Thus far I have confined myself to a consideration of the objective, but the stomach is a sensitive organ, and much valuable work has been done upon its subjective sensations both in health and disease. Experiments on sensation must, almost of necessity, be conducted on the human subject, and so it is not surprising to find that the physician-physiologist has made a larger contribution than the laboratory physiologist to this branch of knowledge. Hurst, in his Goulstonian Lectures on the

"Sensibility of the Alimentary Canal" (1911), showed that the gastric mucosa is insensitive to tactile and thermal stimuli, and that sensations of fulness and pain are due to tension in the plain muscle fibre. The pain of ulcer, formerly thought to be due to chemical irritation of the ulcer base by acid, is now regarded as due to a secondary spasm or increased tension in the muscle fibre. Carlson in an important series of investigations on "The Control of Hunger in Health and Disease" (1916) demonstrated the association of the local hunger-sensation with increased peristaltic activity.

I must now ask you to take a jump which may at first seem quite unwarrantable. Where, you may well ask, is the continuity, when I remind you that in 1849 and 1855 Thomas Addison, a Guy's physician, and the discoverer of the disease of the suprarenal capsules which bears his name, described also a second disease now generally referred to as pernicious, or Addisonian anaemia? It was, in fact, while investigating this anaemia that he happened upon the disease of the suprarenal capsules. In the preface to his monograph on disease of the suprarenal capsules he particularly remarks upon the contributions which pathology may make to physiology, but he could scarcely have foreseen the outcome of his two discoveries.

Pernicious anaemia, a disease characterized by a profound pallor and weakness, sore tongue, splenic enlargement, a tendency in some cases to a mild chronic diarrhoea (which is relieved by the administration of small doses of hydrochloric acid), and in a smaller proportion of cases to degenerative changes in the posterior and lateral columns of the spinal cord, was formerly almost invariably fatal within a few months or years. To-day it can be readily cured. In approximately 100 per cent of cases the test-meal reveals a complete absence of hydrochloric acid. Since the war, largely as the result of investigations by Hurst and his co-workers, the essential unity of pernicious anaemia and a nervous disease known as subacute combined degeneration of the cord (which sometimes develops without conspicuous anaemia) has been demonstrated. In the vast majority of cases in both conditions achlorhydria is present, and histamine provokes no acid secretion.

Prior to ten years ago attempts were made by a variety of means, including the administration of large doses of hydrochloric acid, eradication of oral sepsis and blood transfusion, to cure the disease, but without success. Iron was of no avail. Addison's anaemia remained a baffling problem. Occasional cases, however, in which the gastric secretion returned, have recovered completely. In 1926 as the result of some work by Whipple on experimental anaemia in dogs and the diets which most quickly promoted a regeneration of haemoglobin, it occurred to Minot of Harvard to treat a series of cases of pernicious anaemia with liver. It was soon discovered that, given large quantities of fresh liver, patients with pernicious anaemia could be restored to health and kept well so long as the liver treatment was continued. Later liver extracts were prepared which were equally effective, and in quite recent times, still more potent extracts which can be given by subcutaneous injection, with very rapid regenerative effects on the blood, have been prepared. Extracts from pig's stomach given by the mouth were shown to have a similar potency.

Here then was a substance apparently essential for the maintenance of a normal blood and the absence of which occasioned a disease in persons whose stomachs were unhealthy to the extent that they did not secrete hydrochloric acid. It was at the same time apparent that not all people required liver to keep them well, that vegetarians usually escaped anaemia, and also that deficiency or absence of hydrochloric acid was not itself the cause of the anaemia, since achlorhydria is a not infrequent finding in routine test-meal work and in the absence of any tendency to anaemia.

It next occurred to Castle (1928) to see whether anything could be extracted from meat by normal gastric digestion, which could not be extracted with the known constituents of gastric juice such as hydrochloric acid and pepsin and rennin *in vitro*, and which might provide the anti-anaemic factor shown to be present in liver and some other organ extracts, particularly stomach-extracts. By digesting beef-steak in the stomach of a healthy individual he recovered from the digested products obtained through a stomach-tube a substance which, when fed to a pernicious anaemia patient, caused a reticulocyte response and corrected the anaemia exactly in the same way as liver extracts. This substance is absent from the gastric secretion of patients with pernicious anaemia, but Witts and Hartfall (1933) showed that the gastric juice of patients with a microcytic achlorhydric or iron-deficiency anaemia contained the same property as normal gastric juice.

It now therefore appears that there are two factors, one present in liver and one provided by the stomach which can extract from other foods a factor like the first and as capable of preventing the development of pernicious anaemia and of curing it. These substances are spoken of respectively as the "extrinsic factor" and Castle's "intrinsic factor".

Pernicious anaemia also occurs in a small proportion of cases of gastric cancer and in some cases in which large parts of the stomach have been surgically removed for simple or malignant ulceration. In both we must presume that it is the site of formation of the intrinsic factor which has been removed whether by disease or the knife. In both cases liver can benefit the condition.

Meulengracht (1935), by means of some very careful anatomical, histological and clinical investigations has now shown that in the pig's stomach the pyloric glands and (in the duodenum) Brunner's glands, and to a much less extent the cardiac glands contain the anti-anaemic factor, and when given in dried form to patients with pernicious anaemia produce a good reticulocyte response and correct the anaemia. The fundus or peptic glands on the other hand produce no such effect.

To some extent, therefore, it has been shown that the "lesion" of pernicious anaemia can be localized not only to the stomach but to certain areas of the stomach.

This investigation also helps to explain the fact that the operation of gastrectomy does not always cause pernicious anaemia, Brunner's glands in the duodenum and the cardiac glands being able to supply the deficiency even when a large part of the stomach, including the whole of the pyloric area, has been removed.

Meulengracht's discovery may, perhaps, make it possible to produce pernicious anaemia in animals for the first time, and thereby facilitate future experiment and the testing of therapeutic substances.

In a small proportion of cases both pernicious anaemia and gastric carcinoma are complicated by a peripheral neuritis with weakness and disordered sensation in the limbs. Evidence is accumulating that this can be cured by vitamin B injections. Chronic alcoholism may also be complicated by peripheral neuritis. It was formerly supposed that this was due to some direct intoxication by alcohol. It has recently, however, been shown by Strauss (1934) that it can be cured by liver-extract and vitamin B injections even when the alcoholism is allowed to persist; it is therefore a sequel, in all probability, of damage to the gastric mucosa and so allied to the peripheral neuritis of pernicious anaemia and gastric cancer. Liver itself is rich in vitamin B. May it be that the peripheral neuritis in these three conditions (pernicious anaemia, gastric cancer and alcoholic gastritis) is the same as the peripheral neuritis of beri-beri but due, partly, perhaps, to an insufficient intake of vitamin owing to perturbed digestion and appetite, but partly also to a failure to utilise what is ingested on account of the disease of the gastric mucosa and its secreting cells? At least we have a hint here of yet another continuity, linking gastric physiology with our knowledge of the protective action of vitamins, and promoting the conception of a "conditioned deficiency disease".

Psychology and general medicine as well as physiology owe a debt to Pavlov's work on the digestive glands. Physiology, as well as neurology and general medicine, owe a debt to the recent work of Minot and Murphy, Castle, Meulengracht and Strauss. We have reached a point at which we may claim that the stomach, in addition to its well-established digestive and general nutritional functions and the antiseptic action of its juices, has another very essential and special nutritional function on behalf of the most important of all the tissues of the body—namely the blood and the nerve cells.

It is highly doubtful whether the experimental physiologist could have discovered this for himself or even guessed at the necessity for the discovery. He has not given an animal pernicious anaemia; human cases do not come his way. The links in the chain have been provided by Addison who described pernicious anaemia; by other physicians, including Fenwick, Faber and Hurst, who have insisted on the probable importance of this constant gastric pathology proclaimed by achlorhydria and on the unity of pernicious anaemia and subacute combined degeneration of the cord; by Whipple who treated artificial anaemia in dogs with diets rich in liver and kidney; by Minot and Murphy who treated pernicious anaemia in man with liver; by Castle who devised the human experiment to prove the existence of an intrinsic factor which is not hydrochloric acid or pepsin or rennin; by Meulengracht and his collaborators who have localized the lesion of pernicious anaemia; by the surgeons who have unwittingly caused pernicious anaemia by the operation of gastrectomy; by the physicians who have studied the similar peripheral neuritis of pernicious anaemia, gastric cancer and alcoholism; and by the pathologists and the gastroscopists who have described the morbid changes in the gastric mucosa. In the process a great stimulus has been given to the study of other blood diseases; it has already been shown by Witts and others that another serious form of anaemia differing in certain essentials from pernicious anaemia, although likewise associated with achlorhydria (but not with an absence of Castle's factor), can be cured by massive doses of iron.

Physiology, pathology and therapeutics have joined hands. The clinicians have taught the physiologists things of great importance which they did not know in return for things of great importance which Beaumont, Pavlov, Cannon and others have contributed to medicine. The importance of continuity in thought and action has been repeatedly demonstrated. Facts have replaced theories, truth has dislodged opinion and surmise. The essential unity of the medical and biological sciences in spirit and action can be read in successive chapters.

The earliest discovery of importance was that the stomach secreted hydrochloric acid. The latest discovery of importance is that an absence of hydrochloric acid is sometimes associated with an absence or deficiency of a substance of far greater import to life and health than the acid and other simple constituents of gastric juice. An accident and a disease have between them contributed more to our knowledge of gastric physiology than all the staged experiments put together. On the other hand, physiological experiment with the opaque meal has provided clinical gastro-enterology with its most useful diagnostic method.

Some of us work primarily for the amplification and refinement of natural knowledge by experiment, others primarily for the study of nature's experiments, of disease at first hand, and for the comfort of sick men, women and children, but consciously or unconsciously we are constantly making contributions to the solution of one another's problems. Historical review is valuable if only to remind us of the manner and fruits of this co-operation, and of the necessity for making it more intimate. Observation and experiment have both played their part in the developments which I have all too briefly sketched. It would be invidious to attempt to apportion credit and, indeed, there seems to me to be no reason for trying to do so. We may, however, consider the manner of the respective contributions. Speaking generally we may say that experiment by its measurements and accuracies, by its set conditions and repetitions and comparisons, establishes truth more firmly, proves or disproves hypotheses more certainly, and by its originality claims our constant regard for man's ingenious mind. Observation, on the other hand, by telling us, as accurately as may be, and repeatedly recording what nature is doing, often gives the impetus and direction for experiment and prompts ideas which could not otherwise have been born. By taking the broad view, and by hypotheses and correlations made possible thereby, it counterbalances the more microscopic studies of experiment. From time to time, by utilizing one of nature's accidents (or experiments), as Beaumont and Minot did, it provides the impetus for epoch-making advances in knowledge.

I might have taken the whole field of medicine and considered the advances in the past forty years of surgery, preventive medicine and the specialist branches. I might have described the discovery of Insulin, the important additions to our understanding of the vitamin problem and of endocrinology, or the immeasurable contributions to clinical medicine which have been made in that period by radiology and chemistry. But this would have made a diffuse story and it seemed preferable to confine myself to a more circumscribed subject—even to a single organ—and to consider how our knowledge of its functions and dysfunctions has evolved and influenced our whole thought and practice.

In gastric physiology and pathology, Beaumont, Pavlov, Addison, Cannon, Faber, Hurst, Minot, Castle and Meulengracht, and others too numerous for mention have been members of a team, whose work was separated only by time and place. It is a part of the future duty of medical science, deriving its lesson from history, to diminish the frequency of these separations and so to ensure, whether for philosophic or practical ends, a better and swifter integration of new learning.

VIII. FORTY YEARS OF PARASITOLOGY AND TROPICAL MEDICINE

by

THE LATE
G. H. F. NUTTALL
Professor Emeritus of Biology, Cambridge

PARASITOLOGY IN RELATION TO TROPICAL MEDICINE

THE study of parasitology has received a great impetus during the last forty years through the many problems that have demanded investigation by research workers in medical, veterinary, and agricultural science, helped by the zoologist, botanist, biologist, and chemist. Tropical medicine abounds especially in parasitological problems that concern man's welfare over a great part of the world.

Although there are relatively few strictly tropical diseases caused by parasitic agents, many are the curse of warm countries, while they are infrequent in temperate climates and absent in colder regions. Thus, in the past, malaria occurred in low-lying parts of northern Europe (including England where it has disappeared), but it has continued to be a serious scourge in parts of southern Europe, whilst tropical and subtropical parts of Asia, Africa, and America have always been most seriously affected.

Time does not permit my dwelling on more than two diseases of which I have had personal experience, namely *malaria* (due to a protozoal parasite) and *yellow fever* (due to an ultramicroscopic virus), because they afford striking examples of what well-directed research may accomplish in solving the difficult problems that the medical man has sought to solve.

MALARIA

Malaria affects about one-third of the human population of the world. It is by far the most important disease affecting mankind, causing the greatest mortality, invalidism, and consequent physical incapacity. The disease was known to the Greeks and Romans, who wrote of tertian and quartan fever. The Italians called it malaria, i.e. "bad air", because they believed it was carried in the air, but the cause remained unknown until Alphonse Laveran, a French Army Surgeon, working in a hospital at Constantine, Algeria, in 1880, discovered protozoal parasites within the red blood corpuscles of malaria patients. He studied the parasites carefully, described and figured them. His observations were confirmed in Italy and gradually throughout the world. He thought for long that he was dealing with but one species of parasite.

In Italy, Marchiafava, Celli and Golgi differentiated three distinct species of parasites causing three distinct types of malaria. Golgi discovered that the periodic attacks of fever in tertian and quartan malaria coincide with the cyclical development of the parasites in the blood, young parasites escaping from the infected corpuscles into the fluid part of the blood, say on Monday, Wednesday, Friday and so on in *tertian*, and on Monday, Thursday, Sunday and so on in *quartan*, in simple untreated infections.

Laveran had observed that certain large intracorpuscular parasites, after some minutes upon a microscopic slide, swelled and showed great activity, throwing out lashing filamentous processes which broke away and swam about freely in the blood fluid. This phenomenon was long spoken of as "exflagellation", but its significance was not made clear for years.

The phenomenon of "exflagellation" was regarded by many observers as degenerative until Manson, recalling his work on the worm *Filaria bancrofti* in Amoy, China (1877), advanced the hypothesis that the phenomenon pointed to its being the first step in the "extracorporeal development" of the malaria parasite, for it was only observed in blood soon after its removal from man. In Amoy he discovered that embryos ("*F. sanguinis hominis*") of *F. bancrofti* appeared with a definite periodicity in the circulation, and that when they were sucked up in blood by mosquitoes (*Culex fatigans*) from Chinese patients harbouring the worm, they cast their sheaths in the insect's gut and became highly active. He was so convinced of the significance of the phenomenon of "exflagellation" that he inspired Ronald Ross to work on the lines of his hypothesis, starting with exflagellation and following up the "flagella". It was a brilliant thought and bore fruit. Ross returned to India in 1895 and fell to work. He slaved away overcoming obstacles and interruptions. On 20 August 1897, which he named his "Mosquito Day", he found the pigmented oocysts of the malignant tertian parasites in the gut wall of an anopheline mosquito fed on human malarial blood. He next, in May 1898, attacked bird malaria, due to *Proteosoma*, because Indian patients objected to being experimented upon. He was now on the home stretch at last. He learnt through Manson of McCallum's work (November 1897) in Baltimore, Md., on *Halteridium* (another bird malaria parasite), which demonstrated on the slide that a "flagellum" is the male element destined to impregnate the female element of the malarial parasite. After impregnation the "pigmented sphere", as it was called, became a motile "vermicule". Ross confirmed this on *Proteosoma* and traced the rest of its development in the mosquito, making many fresh preparations and dissections of the insects. He proved that the "vermicules", after penetrating the gut wall, came to rest, grew greatly in size, became filled with many fine fusiform bodies (sporozoites) and burst, thereby liberating the sporozoites which wandered in the insect's body cavity, attained its salivary glands and escaped through the common salivary duct on their wicked way through the mouthparts of the mosquito into the victim they infected. Finally, 22 out of 24 sparrows exposed by Ross to the bites of infected mosquitoes developed bird malaria. Ross's results with bird malaria pointed the way to the Italian investigators Grassi, Bignami and Bastianelli, who traced the same complete cycle of development in the three species of malarial parasites then known as affecting man, illustrating the process admirably and completing the work by infecting volunteers with the corresponding types of malaria.

The most impressive figures relating to malaria in recent years are those from India. A League of Nations Health Organization Report (Geneva, 1932), dealing with the world prevalence of malaria and quinine requirements, gives the population of India (in 1929) as 348 millions, the number of malaria cases, on a moderate estimate, as 100 millions. The grave epidemic of malaria in Ceylon which started in the end of 1934 and ran on into 1935 attacked one and a half millions out of a population of over three millions. About thirteen years ago, an unexampled epidemic of malaria spread northward from southern Russia until it reached Lake Ladoga north of Leningrad and caused vast havoc. These instances will suffice to indicate the gravity of the issue.

In the past it was well known that uncultivated and undrained land, particularly low-lying land in northern countries, was especially prone to be malarious. This also holds for warm climates where, however, malaria may occur some thousands of feet above sea level in ill-drained land, in valleys and along small mountain-side water runs in or near which malaria-bearing anopheline mosquitoes find breeding-places, small or large. Moreover, certain anophelines breed in hollow trees, or water-collecting vegetation. Draining operations have been followed by brilliant results in checking malaria in many places. Drainage of roofs of houses and disposal of refuse demand particular attention. Wire-net mosquito screens to doors and windows of houses are used widely as well as mosquito nets to beds. Such screens need painstaking and constant supervision or they may be a source of danger. Mosquito-control measures alone may suffice in selected localities, but in most cases recourse must be had to chemotherapy.

The bark of Cinchona trees was used as early as 1638 to cure the wife of Count Chinchon of malaria in Peru, of which the Count was Governor. It was afterwards imported into Europe in increasing quantity. The generic name of the trees was established in honour of the Countess, who introduced the remedy. Cinchona trees are now extensively cultivated in different parts of the world, for the purpose of extracting the alkaloid quinine, the long-established sheet-anchor in the treatment of malaria. Enormous amounts are used, but the supply is totally insufficient. For example, in India the amount of quinine distributed was reckoned in 1929 to amount only to 2 grams per head per annum, counting the number of malaria cases at 8–10 millions! Quinine has been improved upon in some ways (by the discovery of synthetic specific drugs like plasmoquine and atebrin), but a better and especially cheaper remedy is clearly in demand. It is hoped that research to this end may be diligently pursued.

YELLOW FEVER

Yellow fever may be defined as an infective, mosquito-borne, non-contagious virus disease, endemic and epidemic, the prominent symptoms in typical cases being albuminuria, jaundice (due to destruction of red blood corpuscles), and haemorrhages from mucous membranes. The name of yellow fever is based on the jaundice, and the Spanish name "vómito negro" is based on the frequency of haemorrhages from the stomach. The degree of albuminuria affords a measure of the severity of a case, whilst anuria is a grave symptom. There are many atypical, mild or masked cases, especially in endemic areas among native populations, where diagnosis is difficult or impossible except by tests recently devised.[1] That naturally acquired immunity, partial or complete, follows recovery from an attack of yellow fever has been known for a long time. Such immunity may persist for life and is of great value to those possessing it because they can sojourn in endemic areas with impunity.

As regards geographical distribution, some of us consider the west coast of Africa as the original home of yellow fever. That Drake in 1585 lost some two or three hundred men of a rapidly fatal disease after leaving Cape Verde Islands for the West Indies, as stated by Carter (1931), is very suggestive, while still more striking is what we know to-day of the wide distribution of the mosquito *Aëdes aegypti*, which occurs across the whole width of Africa from the west coast of the Belgian Congo to Tanganyika, coupled with the results based on research by new methods known as immunity tests to be referred to later.

Trustworthy records of earlier date are few, because of the confusion due to various fevers with some similar symptoms and to imperfect methods of diagnosis. Clear descriptions of yellow fever are recorded from Sene-

[1] I would note here that yellow fever has been repeatedly confused with infectious jaundice and at times with dengue or breakbone fever. Dengue has caused widespread epidemics of recent years in Australia, Texas (5–600,000 cases in a few weeks) and Greece (one-half of the population affected). Dengue is also conveyed by *Aëdes aegypti*, the mosquito which is the notorious vector of yellow fever, though other species have been incriminated as vectors of the latter disease under experimental conditions.

gambia in 1778, from Sierra Leone and the Congo Coast in 1816, and since those dates. In 1935 the disease occurred in Gambia, Gold Coast, Ivory Coast, Dahomey, Nigeria, Togo, and Sierra Leone. In 1927–28 it occurred inland in Nigeria and the Gold Coast.

In America, the endemic distribution of yellow fever has been almost restricted to coastal regions, especially ports of large towns lying in low country by the sea, along great rivers and on the shores of islands in the West Indies. From its endemic centres the disease has extended periodically along shipping routes and reached countries with a temperate climate, mainly the United States, where, in the warm season, it has in the past extended far up the Mississippi, starting at New Orleans. It also extended to large cities such as Baltimore, Philadelphia, and New York. In 1859 it penetrated inland to Texas.

Mosquito surveys made in South America have shown that *Aëdes aegypti* is absent in many parts. This offers a contrast to its wide distribution in Africa and again points to Africa as its original habitat.

The disease is recorded accurately as occurring in the West Indies since 1635, the United States since 1702, Mexico and Central America since 1740, and in South America for the first time as a serious scourge in 1754.

The disease has been brought periodically to Europe, chiefly Spain, since 1800. It has been confined mainly to ports, but in 1878 cases were carried to Madrid. It has entered French ports and visited England, but in neither case did yellow fever become established.

It has caused most deaths in large towns with ports and has been carried in ships from port to port over long distances, for instance, from New Orleans, U.S.A., to Rio de Janeiro. When it reached Lisbon in a ship, observers there declared that some minute living creature on board conveyed the disease, and their suspicions were well founded. Since we know that mosquitoes are vectors of the disease their destruction on shipboard has become a matter of routine, it having been found that they bred in water casks and in the water that accumulates in the bottoms of boats exposed on deck. Precautions are taken with regard to anchorages both as to their distance off shore and position in respect to wind that may blow from shore and convey mosquitoes on board. Mosquitoes may also be carried in trains and motor-cars. It has become abundantly clear latterly that aeroplanes may be a source of danger in carrying insects of various kinds great distances in a short time. Therefore they are being watched both in relation to yellow fever and mosquitoes and to trypanosomiasis (sleeping sickness) in man and its vector the tsetse fly. Regulations (1935–36) now govern aerodromes, aeroplanes and their personnel at places like Juba, Khartoum, and Cairo, and air traffic between Brazil and Africa, etc. Mosquito destruction is practised and other suitable measures enforced.

The mortality, where the disease is endemic, ranges from 25 to 90 per cent among newcomers and 7 to 10 per cent in natives, remembering always that many benign cases occur among natives and are never detected.

It has long been known that persons coming from cold climates are more susceptible than those coming from temperate zones, and that the latter are more susceptible than those coming from warm climates. This is doubtless due to changes of climate and altered mode of life, lowering to a greater degree the natural resistance of persons coming from cold climates.

The deadly effects on newcomers to an endemic centre are well known; thus when I visited the Military Hospital in Havana in 1892 I saw a large ward entitled "il mortuorio" because virtually all the young Spanish soldiers carried into that ward died during the height of the epidemic. At that time Spain lost some 30,000 soldiers from yellow fever alone.

MODERN RESEARCH ON YELLOW FEVER

Modern research on the etiology of yellow fever may be said to have been started by the late Surgeon-General George M. Sternberg (born 1838, retired 1902) of the United States Army, who worked for five years before reaching the conclusion that yellow fever is not due to a bacterial infection. He had had experience of yellow fever as a member of the Yellow Fever Commission from 1879 onwards.

The first to suspect that the mosquito *Aëdes aegypti* conveys yellow fever was Dr Carlos Finlay (1881–1915), a Scotsman practising medicine in Havana when I met him in 1892. He raised the mosquito from the egg and made inconclusive experiments which I criticized adversely in 1899, for they proved nothing. Nevertheless, he was on the right track, as was subsequently proved.

In the Hispano-American war the United States army occupied Havana and tried to better its sanitation over a period of two years, failing utterly in checking yellow fever. A Commission, headed by Dr Walter Reed, working on the disease, excluded all modes of infection that seemed certain or probable, and the Commission finally turned for aid to Finlay, who showed them his suspected mosquito. It seemed possible to the Commission that the mosquito might be a vector in view of the recent discoveries on mosquitoes as vectors of malaria by Ronald Ross in India, followed by Grassi, Bignami and Bastianelli in Italy. Working on the mosquito hypothesis, Reed and his collaborators experimented on volunteers who had not had yellow fever and found that: (1) the virus of yellow fever occurs in the blood of a patient during the first three days of the attack; (2) the mosquito infects itself by sucking the blood during the aforesaid three days; (3) twelve days after the mosquito has sucked infective blood it becomes infective and remains so for 45 days. (Bauer and Hudson (1928), experimenting on monkeys on the Gold Coast, found a single mosquito infective after 91 days.) The Commission, and afterwards Dr Guiteras, experimented on a total of 33 volunteers, of whom 22 developed yellow fever 2–6 days after being bitten by infective mosquitoes. Two of the 22 died, the low mortality being attributable no doubt to "careful nursing", which was drummed into me when a medical student as most important coupled with mild symptomatic treatment.[1]

That the blood of yellow fever patients is only infective during the first three or four days of the attack explains why earlier experience had shown that there was no apparent danger in performing autopsies on the bodies of those that had died of yellow fever.

The disease is due then to an ultramicroscopic virus in the blood, and when infected blood is sucked up by the mosquito there follows a latent period before the insect becomes infective to man.

The recent discovery of susceptible animals which could be used for experimental infection experiments has greatly advanced research on yellow fever because men could no longer be employed even though they served as volunteers. Whilst working on yellow fever on the Gold Coast, Stokes, Bauer and Hudson (1928) discovered that *Rhesus* monkeys from the East Indies were highly susceptible to the disease induced experimentally either by inoculation or through the bites of infected mosquitoes (*Aëdes aegypti*). Of 30 monkeys inoculated 29 died of yellow fever, 2 monkeys were infected by the bites of a single mosquito. Sellards and Hindle found that pieces of liver, derived from infected monkeys on the Gold Coast,

---

[1] In a manuscript letter of a Philadelphia physician, famous in his day, Benjamin Rush (1745–1813), quoted by Osler in 1892, we have an example of the heroic treatment he advised after experiencing the epidemics of 1793 and 1797 in Philadelphia. He wrote: "From a newly arrived Englishman I took 144 ounces, at twelve bleedings, in six days; four were in twenty-four hours. I gave within the course of the same six days nearly 150 grains of calomel, with the usual proportions of jalap and gamboge."

could be transported to England by sea in a refrigerator, the virus remaining fully virulent for 12 days. This has rendered it possible for numerous workers to study the disease in laboratories in Europe and in America. It was found that the virus is capable of traversing intact mucous membranes, and this or slight lesions on the hands may explain why Dr Stokes, in 1927, became infected and died of yellow fever on the Gold Coast when performing autopsies on animals killed during the onset of the experimentally induced disease.

That passive immunity results from injecting the serum of a man who has acquired natural immunity was first demonstrated by Marchoux and Simond (1906) in Brazil. Stokes, Bauer and Hudson (1928) found that 1 c.c. of human immune serum protected *Rhesus* monkeys against infection through inoculation with the virus. That active immunity may be produced by the virus + immune serum has since been demonstrated by independent workers. Several methods of preventive inoculation have been employed and the technique has been improved. In London, Dr Fairley (1936) reports that 951 persons have been vaccinated since November 1932. In January 1934 all whites in Gambia had been inoculated with vaccine from England. Since 1935, some 6000 or more persons have been vaccinated in French West Africa.

Of great practical importance has been the discovery of a method of revealing what may be termed "hidden yellow fever". Mild cases of yellow fever, that cannot be diagnosed, have long been suspected and feared. Masked cases play an important part in the epidemiology of the disease. In 1930, Max Theiler made the important discovery that mice are susceptible to intracerebral inoculation. The yellow fever virus causes encephalitis in mice. This enabled the Staff of the International Health Division of the Rockefeller Foundation to carry out an Immunity Survey with the "protection test" devised by Sawyer and Lloyd (1931). This test was applied in West Africa and South America, using what they termed the "intraperitoneal protection test" in mice subjected to intracerebral inoculation with the virus (details in Sawyer *et al.*, 1932–33, and Sawyer, December 1934). The Commission tested thousands of people, adults and children, finding their blood possessed of immune or non-immune properties, and recorded their results by appropriate symbols placed on maps of North and South America, Africa and southern Europe. Positive indications of immunity covered a much wider area than could have been imagined; thus in Africa positive indications were obtained from Dakar eastward as far as Uganda and the Anglo-Egyptian Sudan. In the Gambia region 20–33 per cent of the native inhabitants were positive. In short, they had at some time in their life had an attack of yellow fever and recovered.

It appears that a damp warm climate is required for yellow fever to be endemic. Falls of temperature in non-endemic areas will cut an epidemic short. There are many records of this. Experiments on the mosquito (*Aëdes aegypti*) have shown that a temperature of 28° C. is the optimum for the development of the virus in the insect and, for the epidemic spread of the disease, the lowest temperature required is 24–25° C. Cold moreover immobilizes the mosquito and breaks the chain of parasitism: virus—mosquito—man. Yellow fever may occur in cold climates if during the warm season yellow fever cases or mosquitoes from endemic centres are introduced.

The history of yellow fever at Panama is especially instructive. The construction of the Canal under de Lesseps began in 1881, and it has been calculated that about 23,000 workmen died before the French enterprise was abandoned. Yellow fever caused many deaths and most of the personnel had malaria. The hospitals were choked with patients and no ways of combating the evil were known, though sulphur was used because it seemed to have been of some service, especially in the United States. The French hospitals were unprovided with mosquito-screened doors and windows, so that the mosquitoes could fly in and out. To keep ants from crawling up the bed posts into the beds the posts were stood in water receptacles which proved ideal breeding-places for the yellow fever bearing mosquitoes in the wards. The female anopheline mosquitoes bred in the vicinity, flew in and out of the doors and windows and stayed mostly in the wards feeding at their list until they flew out to seek the nearest suitable waters on which to oviposit.

Here Vera Cruz may be mentioned because of the former prevalence of yellow fever in that part. I was there in 1885 and passed through the town as quickly as possible in broad daylight to be rowed out to the passenger ship at a considerable distance from the shore, it being deemed very dangerous to linger on the way because of yellow fever.

As the result of the work of Reed's Commission, Gorgas at once started a vigorous anti-mosquito campaign, beginning on 15 February 1901, with the result that *Havana was rendered free from yellow fever in precisely* 80 *days—for the first time in* 400 *years.* I may recall that 35,952 persons had died of yellow fever in Havana during the preceding thirty-seven years, 1863–1900. Guided by this experience in Havana, Gorgas was put in charge of preventive measures at Panama when the Americans decided to build a canal, the French enterprise having been abandoned. The water in which mosquitoes could breed was drained away; water receptacles were screened to prevent mosquitoes from breeding therein. Dwellings and hospitals were provided with wire-netted doors and windows and a rigorous inspection was constantly maintained in connection with all mosquito eradication measures. Starting in June 1904, some cases of yellow fever occurred among the personnel. In January to June 1905 there was a rise in the number of cases of yellow fever (62 cases and 19 deaths in June), and owing to the difficulties that had to be surmounted it was only after 16 months that yellow fever ceased to occur.

The task of sanitation had been carried out at a cost of 1 to 1½ per cent of the total expenditure of building the canal.

I referred previously to the prevalence of yellow fever at Vera Cruz and would mention that anti-mosquito measures were carried out there with brilliant results so that for a period Vera Cruz became a health resort instead of a place that people feared. At a later period the preventive work was neglected and a return of the yellow fever to the town followed. Another case of neglect was that of New Orleans where, in 1905, a fairly severe outbreak of yellow fever occurred through sheer neglect of the precautions that had led to such brilliant results at Havana.

It is interesting to note that there is very much less clinical yellow fever observed in America at present, but further studies are being conducted with immunity tests, and until these are completed we shall not understand the full significance of this apparent subsidence of clinically diagnosable yellow fever.

TRAINING IN TROPICAL MEDICINE

Patrick Manson (1844–1922), after twenty-three years of practice and intermittent research in China, returned in 1889 to Europe and in 1892 became physician to the Seamen's Hospital in London, where cases of tropical disease with which he had been long familiar in the East were frequently admitted. In 1897 he became Medical Adviser to the Colonial Office under Joseph Chamberlain, then Minister for the Colonies. In May 1898, when I first met "Manson of '*Filaria sanguinis hominis*' fame", as I noted in my diary at the time, he was widely known, perhaps even more abroad than at home. In the same year he presided over the first meeting of a Section of Tropical Diseases of the British Medical Association at Edinburgh, and in his opening address did not mince words in laying stress upon the prevailing ignorance among medical men regarding diseases of the tropics, when they left this country for service in the Colonies or elsewhere. He said "there was not one of them who could not pillory himself with the recollection of lives that perished entirely owing to the lack on their part of an elementary knowledge of Tropical Medicine". Strongly supported by Joseph Chamberlain, Manson succeeded in establishing the London School of Tropical Medicine in 1899, gathering an able and enthusiastic Staff about him—it was the first school of the kind. He was opposed by some leading medical men who had no knowledge of the tropics, but he won out and the names of his opponents need not be recalled. The School, like that of Liverpool, served as an example to be followed in other countries having possessions in tropical or semi-tropical parts of the world, and similar schools have been established since in France, Belgium, Germany, Holland, Italy, and the United States.

At the International Congress of Medicine, held in London in 1913, a striking tribute was paid to Manson when Prof. Raphaël Blanchard of Paris, speaking in the Section of Tropical Medicine, rose and acclaimed him as "The Father of Tropical Medicine" in presenting him with a gold medal in the name of an International Committee.

Cambridge having been a pioneer in establishing the first Diploma in Public Health in 1875, Manson and I, with the help of others, appealed to the University and succeeded in having the Diploma of Tropical Medicine and Hygiene established in Cambridge in 1904, in which year we held our first examination with Patrick Manson, Ronald Ross and myself as examiners. We ceased to grant the D.P.H. and the D.T.M. and H. in 1931 because of dwindling numbers of students due to numerous similar diplomas being granted by schools at home and abroad. We set the highest standards we could and consequently our diplomas were the most esteemed. They served a useful purpose by stimulating research and starting a movement which has brought benefit to mankind in many lands.

IX. FORTY YEARS OF EVOLU-
TION THEORY

by

R. C. PUNNETT

Professor of Genetics, Cambridge

EVOLUTION THEORY

IN the minds of almost all who are familiar with his name Charles Darwin stands for the concept of Evolution, for the essential unity pervading the diversity of organic form. Hence in assigning a time limit to the "Darwinian era" we may clearly take as our starting-point 1859, the year which saw the publication of *The Origin of Species*. So long as men believe in evolution, so long in that sense will the Darwinian era continue to be with us. But since for our present purpose some circumscription of its boundaries is necessary I will take as the latter limit the year 1900, the year in which the rediscovery of Mendel's work began to lead to that reorientation of the biological sciences which is still actively going forward. And the question that we have to ask ourselves is—what were the peculiar characteristics of this period of forty years, both in relation to the time that went before and the time that follows after? What were the main ideas prevalent before, during and after the period we are setting out to consider?

Any conception of a process of evolution naturally challenges the problem of causation. These matters were long ago discussed by the ancient Greeks. Into their views I do not propose to enter, but will merely refer those interested to Osborn's *From the Greeks to*

Darwin, where he sums up as follows the results of his delvings in that ancient mine:

The Greeks left the later world face to face with the problem of Causation in three forms: first, whether Intelligent Design is constantly operating in Nature; second, whether Nature is under the operation of natural causes originally implanted by Intelligent Design; and third, whether Nature is under the operation of natural causes due from the beginning to the laws of chance, and containing no evidences of design, even in their origin.

Not for many centuries did any choice among these three views of causation materially trouble mankind. For all their many virtues the Romans appear to have been singularly lacking in curiosity. Though the collecting instinct was by no means wanting among them it seems to have been directed almost exclusively to the acquisition of man-made things such as books, pictures, sculptures and religions. In spite of opportunity there seems to be no evidence of the getting together of anything resembling a museum of natural history objects. Even Pliny, most curious of the Romans, contented himself with the filling of note-books. And the time when curiosity might have led to collection, and this in turn have stimulated classification and inquiry into causes, rapidly passed away; for with the establishing of Christianity a new mental atmosphere possessed the world. The problem of causation was definitely determined in favour of the first of our three alternatives, and men's thoughts were turned to that future life which had been so opportunely called into existence to redress the balance of the present. It is true that a few of the

Church's authorities—notably Augustine and, later, Thomas Aquinas—flirted at times with the second alternative, but in this they had little influence on the great body of the Church and its supporters. Routine marked out the path of the shepherds and the flock ovinely followed. Then came the crash of the Renaissance. Curiosity, so long pent up, surged forth, and natural objects once more became objects of natural inquiry. Contributory were the voyages of exploration now starting, bringing with them new and strange forms of plant and animal life. Collections of natural history objects were brought together, and as they grew in magnitude some system of classification became imperative. In this field the botanists were in advance of the zoologists. Partly this lay in the importance which medical practice attached to plants. The early herbals were compiled from a medical standpoint, but as one succeeded another it became apparent that many plants have points of resemblance to one another entirely unconnected with either their medicinal properties or their importance to agriculture. There gradually grew up the feeling that plants can be arranged in natural groups, and this ultimately led to the production of works in which more and more stress was laid on a natural system, and less and less upon medicinal properties. The animal kingdom, with its vastly greater varieties of form, offered more difficulties. The earliest compendium, Conrad Gesner's *Historia Animalium*, is a queer chaos for the modern reader. Bats are classed among birds, and whales among fishes. Gradually a more and more natural grouping evolved, and by the middle of the eighteenth century was established the famous *Systema Naturae*, a classified conspectus in which Linnaeus arranged animals and plants according to Classes, Orders, Genera and Species. Its acceptance by the scientific world was a formal recognition of a *natural* order in the world of living things. Since the time of Linnaeus classification has undergone enormous development, but without in any way upsetting the conception of such a natural order. When two systematists disagree we should say that both may be wrong, but that only one can be right. Belief in the conception naturally led to speculation as to what lay behind it. Linnaeus and most of those of his time believed that the natural order was the manifestation of the mind of God. The species was as it had issued from the mint of the Creator, though it was conceded that in this evil world slight modifications might at times occur through hybridism or other naughtiness. In other words the species was a constant and immutable thing.

It not infrequently happens that a debatable doctrine may be readily accepted owing to apparent strong support from another which subsequently turns out to be false. Such was the case with the doctrine of the fixity of species. During the whole of the eighteenth century biological opinion was almost entirely under the sway of the curious and erroneous Theory of Preformation. According to it every living thing was brought into existence at the moment of creation, all future generations being preformed and packed away in the germ-cells (for some the ova and for others the spermatozoa) of the original progenitor. Thus for the Ovists the whole of the human race was contained in Eve's ovary, and successive generations were merely an unfolding, or *evolutio*, of the original creation. The doctrine of the fixity of species was a natural corollary of the doctrine of Preformation, and both were heartily endorsed by the Church. For thus the supreme act of creation stood revealed in its true proportions, while at the same time was avoided both troublesome thinking and dangerous speculation.

Most biologists agreed with Linnaeus in accepting the natural system as divinely appointed. Efforts, such as those of Bonnet, to show that living things could be arranged on a progressive scale, with man of course at the head, were made with due reverence to the Almighty, and without thought of any genetic continuity. It is true that here and there voices were raised in an opposite sense, but none was commanding enough to attract a following of any weight. Buffon, popular as he was, vacillated over much, nor was his reputation too good among those who passed for men of science: Erasmus Darwin, though a vigorous and original thinker, was too insular, nor was his vehicle of flowery verse well suited for the purpose: others, such as de Maillet, Robinet and Diderot, were never effective because their speculations, bold as they often were, lacked a sound knowledge of structure on which to base them. For all this while, comparative anatomy was making active progress, culminating in the achievements of Cuvier and his school. Cuvier himself, the founder of palaeontology, was a firm adherent of the theory of special creation, and his influence was immense. Had it not been for him it is possible that Lamarck's views would have attracted more support, and that contemporary copies of the *Philosophie Zoologique* would not be among the rarities of zoological literature.

From Lamarck to Darwin, from the *Philosophie Zoologique* to *The Origin of Species*, is exactly half a century. It may perhaps be termed the Cuvierian era, for it was largely dominated by the concepts of the great French anatomist. For the zoologists it was essentially an age of description and classification, far fuller and more accurate than any hitherto accomplished. Characteristic of it is the production of the great *Règne Animale*, in which neither classification nor sentiment in any way clashed with the basal concept of special creation and the fixity of species. Nevertheless, there were signs that all were not satisfied with the dominant outlook. It was not enough to record living things, however accurately and sumptuously, merely as manifestations of the power and wisdom of the Creator. Might there not be some purpose behind it all? The idea was in the thought of Kant. Organisms are composed of parts which are only comprehensible as conditions for the existence of the whole. The very existence of the whole thus implies an end. Though nature exhibits nothing to us in the way of purpose we can only understand an organism if we regard it *as though* produced under the guidance of thought for an end. Here we are very close to the innate "Perfecting Principle" of Aristotle. Kant's thought influenced Goethe. He regarded a living being as a complex of different elements each referable to a primordial type. Thus, the various parts of the flower are referable to a primordial leaf: thus too are the various segments of the back-boned animals referable to a primordial vertebra. The primordial type becomes modified·in this direction or in that owing to its different position in the series and the different functions this entails. Neither Goethe nor the other "Naturphilosophen" put forward any theory of evolution, but their conception of the transformations of primordial types went some way towards preparing men's minds for such ideas. Nor must we forget that Goethe invented the word *morphology* for a branch of science that was later to play so conspicuous a part in connection with evolutionary theory.

The Cuvierian era was also notable for the advance of embryology. Through the work of v. Baer and others the process of fertilization was coming to be grasped, and the doctrine of Preformation was finally exploded. Very important too was the formulation of v. Baer's *Biogenetic Law*, which, as Singer has pointed out, consists, in effect, of four propositions:

(1) In development, general characters appear before special.

(2) From the more general are developed the less general, and finally the special.

(3) In the course of development an animal of one species diverges continuously from one of another.

(4) A higher animal during development passes through stages which resemble *stages in development* of lower animals.

The formulation of such general principles awakened in these studies an interest which was soon to receive a powerful stimulus.

Meanwhile in another branch of inquiry there was growing up a point of view which was destined to exert great influence in bringing about the Darwinian era. Geologists were shaking themselves free of the "catastrophes" which Cuvier had imposed upon them. In 1830 began to appear Lyell's *Principles of Geology* with its insistence upon the uniformity of geological succession. By abolishing the catastrophe the geologist brought the naturalist face to face with the problem of explaining the connection between the fossil forms of life and those still living. As the science of palaeontology developed, and fresh discoveries were made, it came to be more clearly seen that the distribution of these fossil forms in time accorded well enough with the idea that there existed a genetic continuity between them, while it was not easily to be reconciled with any other hypothesis.

As the Cuvierian era proceeded, the position of the great French naturalist—the orthodoxy of the period— was being steadily undermined. The growth of embryology, of morphology and of palaeontology were telling more and more against the doctrine of the fixity of species, and pointing with increasing emphasis to the existence of a genetic continuity among living things. This involved the conception of the mutability of species through some process of *evolution*, a word first used in this connection by Herbert Spencer in 1852. Indeed the arguments for such a process were marshalled by Robert Chambers in the anonymously published *Vestiges of Creation* which appeared in 1844. Not only does he bring forward those from the sequence of types in palaeontology, from homologies in Vertebrates and from the Biogenetic Law, but he also draws attention to two further lines of argument which Darwin later developed with great effect. Firstly, he points to the existence of the various domesticated races of animals as evidence that certain species, at any rate, are capable of modification, and that such modification can be transmitted: secondly, he adduces the existence of rudimentary organs as an argument against the hypothesis of special creation. Structures such as the small teeth in the foetus of the whalebone whales, the small imperfect additional toes on the splintbones of the horse, the traces of hind limbs in certain snakes, become both intelligible and instructive on the hypothesis of a genetic connection between the different forms of animal life. The *Vestiges* was a widely read book, passing through many editions in a few years. Nevertheless, it cannot be said to have exerted any marked influence on the scientific minds of the time. It did not get the serious consideration which it deserved in spite of obvious shortcomings. The author was not a trained biologist, and as Darwin wrote in a letter to Hooker "his geology is bad and his zoology worse". This doubtless told against him, so that men of such diverse minds as

Sedgwick and Huxley reviewed the book in crushing manner. Nevertheless, Darwin later paid it the tribute of having "done excellent service in this country in calling attention to the subject, in removing prejudice, and in thus preparing the ground for the reception of analogous views". The popularity of the *Vestiges* is evidence that the minds of educated men were not averse to the idea of the mutability of species. Its failure to impress the scientific world, apart from the crudity of its science, probably lay in the fact that it offered no plausible suggestion as to the manner in which the genetic succession of forms could have been brought about. It was the suggestion of such a factor—Natural Selection—by Darwin and Wallace that led to the immediate acceptance of the evolutionary idea. It is true that the publication of *The Origin of Species* excited immense opposition. But that opposition came almost entirely from outside the ranks of science, and the arguments used were largely of the nature of appeals to existing prejudice, particularly where the status of man himself was concerned. The immediate support for the new views forthcoming from Lyell, Hooker and Huxley led rapidly to acceptance by the scientific world. The need of some synthesizing idea bringing into relation with one another the great body of facts derived from palaeontology, morphology and embryology had long been felt, but their synthesis through the idea of evolution had been delayed through ignorance of any factor by which such evolution could be supposed to have been brought about. That factor was now supplied, and botanists and zoologists at once set out to rearrange their facts and systems in the light of an evolutionary succession of life. For several decades their main interest lay in the construction of genealogies relating together the great groups of the animal and vegetable kingdoms. Among animals three of these great groups—Cuvier's Vertebrata, Mollusca and Articulata—were obviously pretty homogeneous, and within them the refined study of adult morphology and palaeontology was largely relied on. It was otherwise with Cuvier's 4th Embranchement of Radiata, and of the miscellaneous class Annelides. Small, and often insignificant forms, they had been far less studied. But with the illuminating idea of evolution they offered a scarcely explored field where facts of the highest interest might readily turn up. It was among these lowly creatures that the evolutionary morphologist might expect to find forms of life which shed light upon the relations of the great homogeneous groups with one another. He turned to the task with enthusiasm, and with the aid of the microtome and of vast improvements in microscopic technique was produced the prodigious volume of morphological research which crowds the journals of that time. But perhaps the most characteristic type of research of the period is that known as Comparative Embryology. The idea upon which such work was based arose from the impact of the evolutionary idea upon the Biogenetic Law. The result was the Recapitulation theory, which stated that the ontogenetic history of the individual was a repetition, often blurred and abbreviated, of the phylogenetic history of the group to which it belonged. It received an immense stimulus from the brilliant work of Kowalewsky on the Tunicata, or sea-squirts, a group hitherto placed among the Mollusca. Through the study of their embryology Kowalewsky was able to show that in the course of their development they pass through stages conforming to a primitive vertebrate plan, with a free-swimming tadpole-like larva. This work led to a reconstruction of the genealogical tree of the great vertebrate phylum, taking its origin back to a simple hypothetical form without any vertebrae at all. A few years later Bateson showed that the peculiar worm *Balanoglossus* passes through morpho-logically comparable stages and should therefore be regarded as having sprung earlier from the same stem as that which later gave rise to the Tunicata, to *Amphioxus*, and to the Vertebrates. Further, it was known that other members of the *Balanoglossus* group passed through an apparently quite different life history with a free-swimming larva resembling that found in certain Echinoderms. From careful morphological investiga-tion, guided by generalizations such as the germ-layer and the coelom theories derived from other forms, it was shown that the two kinds of development could be reconciled with one another. So was the conclusion reached that mammals and starfishes had sprung from the same primitive stock, though separating very early from each other and pursuing very different evolutionary paths. What Comparative Embryology was doing for the Vertebrates it was doing also for every other group in the animal kingdom, and it was not long before an

enthusiast such as Haeckel could confidently state the
evolutionary history of any form of life, however complex,
in its lengthy passage from primordial slime. In the
pursuit of these ideal genealogies there was no check
upon the imagination: for the only branch of science
that could have supplied it, viz. palaeontology, was from
the very nature of the case unable to do so. These far-
away small soft things could have left no record in the
ancient rocks. So through the first few decades of the
Darwinian era the Comparative Embryologist passed
from triumph to triumph. Balfour's *Comparative Embryology* issued in 1880–81 is a landmark in these studies, and
the traditions there established were carried on by Ray
Lankester in England and by Korschelt and Heider in
Germany.

Darwin's great book was entitled *The Origin of Species*.
Its thesis amounted to a denial of the existence of species
as hitherto understood. What we term a species is
merely a time concept—a cross-section at a given
moment through a gradually changing and genetically
connected series of life forms. Though the change may
be imperceptible to our appreciation a species is different
from what it was yesterday, and will be different again
to-morrow. With the doctrine of the continuity of
genetical succession it was deprived of that character
of fixity with which it had been endowed by the theory
of special creation. Curiously enough, as Bateson
pointed out, the systematists, the very people on whom
the new doctrine might have been expected to exert the
greatest influence, stood obstinately aloof. Though they

might pay lip homage to evolution, they continued unmoved to discuss what was and what was not a "valid" species. The men whose work brought them into closest contact with species treated them as realities, and with the conviction of some degree, at any rate, of fixity about them.

Thirty-five years after the *Origin* had appeared a new note was struck when, in 1894, Bateson published his *Materials for the Study of Variation*. Darwin's theory had been based in part upon what was termed the Principle of Variation. No two living things were exactly alike and their differences were what Natural Selection worked upon to bring about the gradual change of form we term evolution. In the absence of exact knowledge the phylo-genetic architect had assumed the existence of unlimited variation in all directions for the construction of his genealogical edifices. Bateson protested that after all the nature of species had *not* been solved; that the forms of living things remained "specific" in spite of the doctrine of continuity in genetical succession. Some other means of attacking the age-old problem must be found, and he suggested that the fairest hope of success lay in the study of the nature of variation itself. We must discover what kinds of variation do actually occur in Nature, for these are the limitations by which evolu-tionary change must be bounded. From his collection of facts he showed that variational change is in large measure *discontinuous*—that in a freely interbreeding community may occur definite varieties without the existence of any series of intermediate forms to link them with the normal. May not this demonstrable discontinuity in variation lie at the root of the discontinuity which is so characteristic of closely allied species? Bateson's call to a new study excited but little interest, and in a few years his book was to be found among the list of cheap remainders. Orthodox complacency required a further stimulus, and this was soon to be provided in de Vries's *Mutations Theorie*. The publication of this work in 1901, coinciding as it did with the rediscovery of Mendel's paper, may be said to mark the close of the Darwinian and the opening of the Mendelian era.

We must now turn to another branch of inquiry which, at first seemingly independent, is now playing a part of the highest importance on the question of the nature of species that Darwin had brought into such prominence. Following Schleiden's historic paper of 1838 the cell doctrine became firmly established by the work of Kölliker and others during the next two decades. Improvements in microscopic technique led to rapid development in our knowledge of the nucleus, and by the middle of the Darwinian era, chiefly through the researches of Strasburger and Flemming, the complex processes of cell division were unravelled, and the chromosome stood ready to assume the important rôle that was later to be thrust upon it. It was some time, however, before it was drawn into the orbit of the species problem. Beyond the fact that their number was constant for all the cells of a given species little was yet heard of chromosomes in this connection. It may, however, be mentioned that this more precise knowledge of the mechanics of cell division now available served to place a curb on those speculations which attempted to analyse the cell into smaller hypothetical units in attempts to explain the phenomena of heredity. One cannot imagine Darwin putting forward his theory of "gemmules" had he been acquainted with the nature of cell division.

The Darwinian theory of evolution through natural selection had been built upon the twin pillars of Variation and Heredity. The continuous nature of variation, so essential to the theory, was, as has been pointed out, challenged by Bateson in 1894. We may now turn for a moment to heredity. Darwin frankly confessed that very little was known about the subject. Speaking generally he and his followers subscribed to the popular saying that "like begets like", though with some reservations. The general rule might be upset through the operation of some unknown factor to produce *reversion*, where progeny "throw back" to some remoter ancestor instead of closely resembling their parents. Again, the unaccountable "sport" that might at times arise was generally referred to some abnormal environment, such as domestication. But the "like" begotten by "like" is never exactly alike. There was supposedly some range in variation upon which natural selection could work. It was all rather vague, but sufficiently in keeping with popular opinion to excite little criticism on the part of the scientific, whose ignorance in these matters was on a par with that of the crowd. The stock-breeder and the fancier could have told another story, but with the publication of *Animals and Plants* a few years after the *Origin* the scientific world ceased to take any interest in their observations. The matter was regarded as closed for, as a distinguished zoologist remarked, "Darwin had swept the board". It is a curious fact that the publication of Darwin's work, instead of stimulating, should have killed all interest in breeding research. Nevertheless, an effort on quite different lines was made to clarify ideas upon heredity. Originating in the fertile mind of Francis Galton it consisted essentially in the application of statistical methods to the study of resemblances between groups of relatives in different degree. On the basis of specially constructed correlation tables for this or that chosen character Galton formulated the so-called "Law of Ancestral Heredity" which stated the *average* contribution made to an individual by parents, grandparents and remoter ancestors. This essentially quantitative view of the subject was enthusiastically taken up by the mathematician, Karl Pearson, and at the close of the century led to the foundation of the Biometrical School. Based upon the view that all variation is continuous and equally transmissible, it rapidly collapsed when Mendelism proved these views to be untenable.

One further attempt at understanding the nature of heredity should be mentioned here. This was August Weismann's famous germ-plasm theory. Though entirely speculative it was the product of a trained biologist who in earlier years had made important contributions to science. His theory emphasized the continuity of the germ-plasm through successive generations. The body, or soma, was merely a temporary structure formed by the germ-plasm for its own conservation. Their function of conserving and passing on the germ-plasm fulfilled, bodies wore out and died, but the germ-plasm was immortal. The form taken by the soma depended primarily upon the germ-plasm which formed it, and though it might be altered by external conditions during its lifetime such alterations could not be transmitted. There was no room in the theory for the transmission of somatic alterations in the Lamarckian sense, and Weismann utterly rejected them. By demanding evidence and finding none that would stand criticism he did a great service to biology. But it must be confessed that in so doing he tended to weaken the Darwinian position. Although Darwin himself regarded natural selection as the main factor in producing evolutionary change he attributed much to the influence of the environment and to the Lamarckian factor of "use and disuse". Weismann swept all this away and saw in natural selection the only factor for bringing about evolutionary change. Variation for him was initiated in the germ-plasm—partly through *amphimixis*—the mingling of two germ-plasms that occurs at every act of fertilization—and partly through a selective struggle in the germ-plasm itself among the complex system of "biophors, ids and idants" of which it was composed. The fact that Weismann localized his system of hereditary units in

the chromosome and so brought his theory into con-
sonance with the growing science of cytology had much
to do with the influence that Weismannism undoubtedly
exerted. But by cutting away the Lamarckian and
environmental props and by insisting upon the "omnipotence" of natural selection in evolutionary change
he threw upon this last factor a burden which, in the
opinion of many, was too great for it to bear.

Darwin was a philosopher, but at the same time he
was a naturalist—one of the greatest. In the former
capacity he wielded enormous influence over the
laboratory, by stimulating those researches into Comparative Embryology which may be said to strike the
dominant note of the era. As a naturalist also his
influence on the period was very considerable, and
principally in the study of what has been termed
Adaptation. That animals and plants are on the whole
peculiarly adapted to the circumstances in which they
live had long been recognized. The pious who were also
naturalists had found in these adaptations further
evidence for the power and beneficence of the Creator
who in the beginning had taken thought for all these
things. Such was the line adopted by John Ray in his
work on *The Wisdom of the Creator* which appeared in
1691. The promulgation of the idea of natural selection
at once vested these inquiries with fresh interest. On
the Darwinian view any kind of variation which gave
to its possessor even a slight advantage in the struggle
for existence would gradually accumulate as the generations passed. Taking natural selection for granted one

must suppose that the various characters of animals and plants have come to be what they are because they are *useful* to the individual in the struggle for existence. Hence if a character can be shown to be useful its very existence bears witness to the efficacy of natural selection. The argument is somewhat circular, and I am not defending it. But it is the sort of argument that was at the back of the minds of those who turned to the study of adaptation. In his books on *The Forms of Flowers* and on *The Fertilisation of Orchids* Darwin himself had blazed the trail. As the result of numerous experiments with plants he had come to the conclusion that self-fertilization was injurious, a thing abhorred by Nature. As the result of numerous observations he had noticed that many flowers are so constituted as to be more readily fertilized by foreign pollen arriving through insect agency than by the pollen they themselves produce. By allowing of, and often even insisting on, cross fertilization, the form of the flower must be regarded as an adaptation beneficial to the species. For the progeny so arising would be more vigorous and better able to withstand competition. Such a useful adaptation might well be brought about through the operation of natural selection. The study of Adaptation offered a pleasant alternative to the discipline of the laboratory, and many were those attracted to it. On the animal side were recorded innumerable cases of "protective resemblance"—a butterfly resembling a leaf, a spider simulating an excrement, and so forth. In most cases the description wound up with a few remarks on the efficacy of natural selection. In Ray's time it had been the Omnipotence of the Creator. Very seldom was any effort made to test whether the case described, e.g. a variation in colour, *was* of any value to its possessor; whether those in possession of the presumably advantageous variation *actually did* have any advantage as judged by survival value. And that of course is what is wanted.

Among the phenomena of Adaptation none has excited greater interest than that of Mimicry. In its special sense the term is applied to cases where a species, generally an insect, resembles in appearance another species which may belong to a different family, or even order. Such resemblances were first noticed by Bates when collecting butterflies in South America. The *Origin* had just been published, and in 1862 Bates formulated a theory to account for these resemblances in terms of natural selection. He had observed that in the cases he had come across one of the forms, and that the more plentiful, was not attacked by birds, and this he attributed to its possessing a disagreeable flavour. It was also characterized by a conspicuous colour pattern. This was supposed to have developed through the operation of natural selection, for it was of advantage to its possessor in advertising its unpleasant properties. After a trial or two the would-be predator would associate the striking pattern with an unpleasant flavour, and would henceforth leave it alone. The more striking the pattern the more efficient the advertisement—a point which was seen to by natural selection. The development of such a striking pattern, or "warning coloration", was the first stage in the production of an example of mimicry. Bates observed that the other species concerned belonged to a group which might be preyed upon by birds, since it had not been able to develop a nauseous flavour, even through the operation of natural selection. Careless in this respect natural selection came to its rescue in another way. By favouring those variations in which it approached the pattern of the nauseous "model" it gradually built up a similar pattern in the palatable "mimic". Henceforward the mimic enjoyed immunity from the attacks of the postulated predator, which had learned to associate the pattern with unpleasant qualities. The argument was accepted by Darwin, who devotes several pages to these mimetic resemblances in the fourth and subsequent editions of the *Origin*. Later, Wallace and Trimen recorded further instances of mimicry among Oriental and Ethiopian butterflies, and since that time large numbers of cases have been described, chiefly from tropical countries. Since its discovery mimicry has always been regarded as of the very first importance by the advocates of natural selection as *the* factor in bringing about evolutionary change. Weismann regarded it as the keystone of the arch. As he and those who think with him would say—Here we have all these peculiar cases of special resemblance, as to the reality of which there can be no doubt; on the theory of natural selection alone are they comprehensible, and no other plausible explanation has ever been adduced. Throughout the

Darwinian era and beyond it the subject of mimicry excited the keenest interest. Around it there has grown up a voluminous literature, though this, as Singer justly remarks, is peculiarly naïve and unscientific. For the critical mind there are many reasons for dissatisfaction with the theory as it stands. Much that it postulates is utterly at variance with the established results of genetical research. Yet the existence of these peculiar cases of resemblance remains one of the strongest and most fascinating problems of the naturalist. Were it to be solved much else would be solved also. But the solution is still to seek.

Wallace, as we have already seen, was one of the pioneers in developing the theory of mimicry. He was also a pioneer in another line of study with an intimate bearing on the doctrine of evolution—the study of the geographical distribution of living forms. Darwin himself had earlier drawn attention to the peculiar nature of the fauna and flora of certain islands. In *The Voyage of the Beagle*, when discussing the collections he had made in the Galapagos Islands he pointed out that not only were many of the species peculiar to the group, but that each individual island contained forms which were not present on the others. Yet, though distinct, they were not very different from one another. Why had it been necessary to create all these slightly differentiated and narrowly distributed species? That one species should have been created for each small island seemed hardly rational; but how then had these different species arisen and why did they belong to South American genera?

It was largely this problem of island forms, as propounded by those of the Galapagos Archipelago, that set Darwin's thoughts travelling to the problem of the formation of species. With the formulation of his theory, this group of facts, like so many others, seemed to find a rational explanation. The subject was later developed by Wallace in greater detail. He evolved a geographical system for the distribution of animals which has stood the test of time. The nature of the fauna of an area is largely wrapped up with the geographical changes that the area has undergone. If these changes are taken into account, and if we assume an organic evolution from more primitive to more specialized forms, then the geographical distribution of animals becomes intelligible. New Zealand contains no indigenous terrestrial mammals because these had not yet made their appearance when geological conditions led to the isolation of that country from continental land. Australia lacks indigenous placental mammals because it received its mammalian population when the more primitive marsupial alone existed. Before the placental evolved elsewhere Australia had become cut off from the rest of the world. Provided that the postulated changes in land areas are confirmed by the geologist, and that the phylogeny assumed agrees with that of the morphologist, the fact that an explanation of the past and present distribution of animals is possible is definitely in accordance with the theory of organic evolution. That both geologist and morphologist have in the main concurred is strong circumstantial evidence for the theory. As to the manner in which this evolution has come about it offers no evidence, and for this reason perhaps, it remains, in spite of accumulated data, in much the same position philosophically as that in which Wallace left it. But it is not unlikely that fresh interest will develop in it when we are in the possession of more precise knowledge of the manner in which variations arise.

Most of the lines of research actively pursued in the Darwinian era—morphology, comparative embryology, palaeontology, geographical distribution, adaptation—were lines of inquiry directly stimulated and largely remodelled as a consequence of the Darwinian doctrine. Others like cytology lay outside its orbit, and to these we may add experimental embryology, or, as some prefer to call it, developmental mechanics. Initiated by Wilhelm Roux in the 'eighties the new line of study was definitely experimental, seeking by such means to understand the forces at work in the embryo. Hitherto the embryologist had been content to describe the normal course of development, and to draw from it phylogenetic inferences. For the newer school phylogenetic problems presented no interest. It was the aspect of organization that attracted them, and they sought to throw light upon the causes of development by studying it under abnormal conditions, such as mutilation, distortion, changes of chemical or physical environment, etc. In the hands of Driesch and others, especially Jacques Loeb whose experiments on artificial parthenogenesis excited much interest, this new line developed rapidly, particularly in America. Historically it is of interest in that its exponents broke clean away from the Darwinian tradition. Hitherto, under the denomination of the Biogenetic Law, the *cause* of developmental sequence, in so far as embryologists concerned themselves with such things, lay in phylogenetic history. Many Crustaceans developed a nauplius larva at some period of their ontogeny because, in the remote past, they had all descended from a nauplius-like ancestor. When the nauplius larva failed to materialize it was because the developmental sequence had been abbreviated through natural selection; and in such cases it was the self-appointed task of the embryologist to detect this masked nauplius stage and so to relate the abbreviated to the fuller sequence. And there the matter ended. The aim of the developmental mechanists was to inquire how far embryological sequence could be interpreted on purely physical grounds. In one respect, however, the new line had considerable influence on Darwinian studies. By opening up a fresh and independent avenue of research it attracted away the rising generation of embryologists, and the deflection of ideas and aims contributed largely to the wane of morphology towards the end of the nineteenth century.

And now arose another line of study by which the Darwinian doctrine was to be tested. I refer to the experimental study of heredity which suddenly came to the fore with the rediscovery of Mendel's forgotten paper in 1900. Though, properly speaking, this movement lies just outside the Darwinian era as defined above, it is in some ways so intimately associated with

it that I may perhaps be pardoned if I speak briefly of
its impact on the older doctrine. For the evaluation of
any doctrine we ought, if possible, to consider it in
relation to what came after as well as to what went
before.

We have already seen how, in 1894, Bateson had
challenged the idea of continuous variation implicit
in the Darwinian teaching; how he had pointed out
that variation might be discontinuous—that two well-
marked variants might be found among a freely breeding
population without the existence of those intermediate
forms which, on the accepted doctrine, should occur.
Moreover, he had suggested that such discontinuity in
variation might be at the root of the discontinuity of
species. Might not species be real and permanent
things sharply marked off from one another even as
the systematist averred, and not mere cross-sections of
a gradually changing life-sequence as the Darwinian
theory supposed. Convinced of the existence of discontinuity in variation Bateson set in train experiments
to discover how such variations behaved in the hereditary
process. He had not, however, proceeded far when the
news of Mendel's work reached him. Discontinuity in
Variation had its counterpart in the Discontinuity of
Heredity. This brought with it the conception of definite
entities—factors they were then called—in the germ-cell
corresponding to definite characters in the individual.
Moreover, these factors were conceived of as, in general,
permanent things passing unchanged from gamete to
zygote and from zygote to gamete for innumerable

generations. Upon them depended the characteristics of living things. The claim was no longer that variation *might* be discontinuous, but that all heritable variation *was* in its nature essentially discontinuous. Such a claim was at once resisted by the Biometricians, for they saw that its acceptance knocked from under them the foundation upon which they were building. The controversy that ensued was short but decisive. The first impact of Mendelism on the Darwinian doctrine was to discredit the current view of the nature of variation. Begun under these auspices the experimental study of heredity went rapidly ahead. Its adherents were too actively engaged to pay much attention to relating the new knowledge to evolutionary doctrine, though incidentally the phenomenon of reversion on crossing, which had so puzzled Darwin, received an explanation. Attention was largely focused on the problems of linkage and of sex. So went by a decade, and then came into the picture the little fruit-fly *Drosophila*, and with it came the *gene* theory in which the *factor* of earlier workers was re-christened in good American and assigned a definite habitation in the chromosome. Research has now made it certain that the genes, upon which depend the manifestation of characters in the organism, are bodies of a size calculable within reasonably narrow limits and are arranged within linear series in the chromosomes. The material basis of heredity was laid bare, and the gene with its predictable behaviour and ascertained position in a visible structure has taken the place of the purely hypothetical gemmule, pangen or biophor of earlier speculation. Thus was established that intimate contact between genetics and cytology which is now the feature of modern work. It is unusual to find two branches of study begun, and for some time pursued, entirely independently, fuse together as it were into a single whole. Yet to-day neither cytology nor genetics can be profitably studied apart. A peculiarity in chromosomal arrangement is at once reflected in some peculiarity in heredity, and, conversely, seemingly abnormal transmission at once leads to an examination of the chromosomal complex. Quite recently the pioneer work of Muller in exposing *Drosophila* to the influence of X-rays has demonstrated that not only can fresh and unknown mutations be artificially produced, but that through genetical analysis the nature of the chromosomal changes involved can be discovered. As might be expected the new technique is producing its most striking results in the more easily worked plant with its simpler structure. Notably in *Datura*, *Crepis* and *Nicotiana*, genera already largely analysed from the genetical standpoint, have the results been most revolutionary. For it has been found possible to produce new forms of plant life, with definite characters of their own, breeding true to these characters, and at the same time showing sterility towards the parental forms. In other words there have been created new species, the essential nature of which, as compared with the parental forms, is a rearrangement of chromatic material, a realignment of the genes, accompanied perhaps by losses or additions of some or others among them. Although this stage has not yet been reached in animals the evidence from the study of the chromatic material in closely related forms of *Drosophila* points in the same direction.

That all this must necessarily lead to a recasting of our ideas concerning the nature of species and their mode of origin is obvious; for to-day we are in possession of vital knowledge which was lacking in the Darwinian era. Understanding, as we do, much of the process of heredity and something of the nature of variation we view with altered vision the age-old problem of species. That their existence is based on some mutational process we are convinced, and we know too that there is every prospect of gaining an insight into that process in the laboratory. But at present we do not know for certain whether this process is necessarily *extrinsic*, whether mutations are always due to the action on the germ-cells of some force external to the organism. There is always the possibility that there may be some form of *intrinsic* mutation, dependent upon some part of the highly complex cell mechanism failing to keep step, as it were, over a long stretch of countless cell divisions, and thereby leading to the formation of something fresh, should it prove to be viable. Mere speculation as this is, it may serve to emphasize that we are as yet in no firm position to hazard conjectures on the origin of species in spite of the fact that we have learned a great deal about their nature.

Meanwhile, in concluding, we may take a brief backward glance over our allotted period. What have we taken from it, and what have we discarded? In the first place we still hold by the theory of evolution, regarding the world of living things as a dynamic, and not a static concern. The idea is of course an old one, and Darwin's chief glory, as Butler said, is not that he discovered it but that he made men believe in it—and what glory, he added, could be greater than this? Natural Selection also is still with us, but in a rather different sense. For Darwin and his followers, owing to their conception of the nature of variation, it was in large measure a *creative* force, accumulating small variations until they attained a magnitude that enabled them to play a part in specific change. Our insight into the nature of variation has changed all this. Natural Selection, said Bateson, is a true phenomenon, but its function is to select. It plays the part merely of a selective agent on heritable variations, which have already arisen through an independent process of mutation, conserving the beneficial and rejecting the inimical, while producing no effect upon those that are neither the one nor the other. Through such limitations of its scope we are released to-day from the necessity of finding a use in everything merely because it exists. On the other hand continuity in heritable variation has gone, and with it the idea of continuity between species. Species are once more sharply marked off things with hard outlines, and we are faced once more with the problem of their origin as such. The idea of yesterday has become the illusion of to-day; to-day's idea may become the illusion of to-morrow. "For", says Meredith, "the mastery of an event lasteth among men the space of one cycle of years, and after that a fresh illusion springeth to befool mankind." Doubtless many masters of the event will follow after Darwin and Bateson in wielding the Sword of Aklis, and through the dispelling of illusion after illusion mankind may eventually encounter the ultimate residue, perhaps the ultimate of all illusions, which we optimistically designate as truth.

X. FORTY YEARS OF GENETICS

by

J. B. S. HALDANE

Professor of Biometry, University College, London, formerly
Sir Wm. Dunn Reader in Biochemistry, Cambridge

GENETICS

IT is difficult to speak on the history of one's own time, for two reasons; first, that one cannot see the wood for the trees—one is overwhelmed by details, and one does not know which details will be regarded by posterity as important. The second reason is that it is perhaps easier to speak ill of the dead than the living, in spite of proverbs to the contrary. I must therefore crave your indulgence if much of what I say will appear to you to be rather trivial. That is because of my closeness to what I am describing; and if I do not pass sufficiently final judgment on some living persons, that is again because one is right, I think, to wait for the verdict of posterity.

Before we begin the history of the last forty years, it will be well to say a few words about the position in 1895. At that time Darwin's views were very generally accepted by biologists; such opposition to them as existed was based rather on religious grounds than on a study of scientific facts. The reasons for this remarkably rapid and widespread acceptance of Darwinism were, I think, as follows. Darwin's theory explained a very large number of facts of palaeontology, of comparative anatomy, of comparative embryology, and of geographical distribution. It enabled further facts to be predicted. The predictions in those fields were soon verified, and they have been on the whole verified during the last forty years. At that time biologists were mainly occupied with problems such as these, and they did not realize that while Darwin's account of the historical facts of evolution might be entirely correct, the causal explanation which he gave for them was very much more doubtful.

The broad historical features of evolution are not, I think, now in serious question, but there is still considerable doubt as to how they should be explained. Similarly, the rise and fall of human civilizations, which are given facts for history, may be explained in terms of economics, of biology, of the rise and fall of races, or of the influence of new ideas on human beings, and so on.

Now Darwin's account of variation in animals and plants was as admirable as his causal analysis of it was unsatisfactory. He took the view that variations of all kinds might be inherited, and therefore he accepted Lamarck's view, which had been held by most of his predecessors, that the effects of use and disuse on organs are inherited. He did not, like Lamarck, regard use and disuse as accounting for evolution. He insisted on the necessity of natural selection as well. That view was first systematically challenged by Weissmann, and in 1895 the controversy between the followers of Weissmann, who disbelieved in the inheritance of acquired characters, and their opponents, was exceedingly vigorous. Now in the course of the nineteenth century two profoundly important pieces of genetical work had been done which had not been incorporated into the general system of biological ideas. The two workers responsible were de Vilmorin and Mendel. R. L. de Vilmorin was a member of a family of French seedsmen, who have carried on the firm in the family since the late eighteenth century and are doing so still. They were responsible among other things for the production of the modern sugar beet in the early nineteenth century. De Vilmorin's principle was as follows: in a highly inbred line of plants you will not get your best results by selecting the best individuals. You will do better to select a line on the basis of its average performance. That in itself does not seem very revolutionary, but if you follow out its implications you will see that they are entirely opposed to the Darwinian idea that all kinds of small variations are inherited. In its modern form, due mainly to Johannsen, we can state that when a pure line of plants or animals is established by inbreeding or other methods, the small differences which occur between members of a pure line are in general not inherited at all. We have got, as it were, a zero of heredity. We have got in the pure line a set of organisms which are genetically homogeneous. When we are studying a complex phenomenon it is comparatively important to eliminate our variables one by one. If we are studying the volume of a gas we must first keep its temperature constant to discover Boyle's Law, and secondly keep its pressure constant to discover Charles' Law. If we wander about with a variety of temperatures and pressures it will be much longer before we discover anything. De Vilmorin discovered how to eliminate the hereditary element in variation. The importance of his work was not realized till the twentieth century.

The second neglected worker was, of course, Gregor Mendel. We may sum up Mendel's work by saying that he introduced atomism into genetics. He found that many of the differences between pea plants were due to factors, or, as they are now called, genes, which behave as units and are reproduced generally unaltered from one generation to another. Just before our period opens, in 1895, W. Bateson, of St John's College, had published his very important book, *Materials for the Study of Variation*. He placed emphasis on the discontinuity of variation, whereas orthodox Darwinian theory laid stress on the continuity. Galton had also recently published The Law of Ancestral Heredity, a statistical conception mainly based on the colour inheritance in basset hounds. It has proved to be of less importance than was anticipated at the time, although Galton's studies on frequency curves have had an enormous influence on subsequent work. Finally, one must add that Weissmann had just given the first satisfactory account of meiosis and fertilization. He pointed out that the number of chromosomes halved at meiosis and doubled again at fertilization.

Ten years later, in 1905, the picture was entirely changed. Instead of arguments about the complete exactitude of the Pentateuch there was a considerable group of facts which did not agree too well with any of the pre-existing theories. What had accounted for this sudden change? It is very difficult to give a complete analysis, but I think we can point out some of the reasons. In the early nineteenth century a good deal of very fine work on plant breeding was done, not to any appreciable extent by professional botanists. The mid-nineteenth century, on the other hand, was probably the age when systematic biology and botany assumed their greatest economic importance. It was an age of rapid exploitation of newly discovered countries; and this made a knowledge of species, their character and geographical distribution, exceedingly important. But before the end of the nineteenth century, the economic situation had changed. For example, Canada was becoming important as a source of wheat rather than as a source of furs. Scientific plant breeding was being started on a very large scale. Thus, in 1892, Saunders in Canada made the first cross which led to the production of Marquis wheat in 1904, an event which extended the wheat area a long way to the north, although Marquis wheat was only first distributed in the year 1909, seventeen years after the first crosses were made. It is not, I think, a mere coincidence that the year 1900 saw not only the rediscovery of Mendel's work, but the first experiments by the Home Grown Wheat Committee of the National Association of British and Irish Millers to test Canadian wheats in England. Similarly in Holland, Prof. Broekema had crossed the English Squarehead's Master and Dutch Zeeuwsche in 1885. From this, Queen Wilhelmina, one of the most popular wheats in Europe, arose.

Besides this growing interest in plant breeding due to sound economic causes, there were of course intellectual reasons operating in academic circles. It took about a generation for biology to recover from the shock of Darwinism just as it took considerably longer for physics to recover from the shock of Newton; and it was only after a generation that people began to ask where the real intellectual gaps in Darwinism lay. There was so much that was theologically startling and philosophically puzzling in the descent of man. Another reason for the revival of interest in genetics lay in the Eugenics movement. This movement, I think, must be regarded largely as a product of the class struggle based on the desire of the governing class to prove their innate superiority. It led to renewed investigation of the problems of human heredity, much of which was of very great value.

Our first date, I think, is 1897. In that year Bateson, with the aid of a grant from the Evolution Committee of the Royal Society, hired an allotment near the Botanical Gardens in Cambridge and began his classical experiments on poultry breeding and plant breeding. You will notice that he got his money from the Evolution Committee. The idea was that by studying the inheritance on the various types of comb-structure, feathers, and so on, in poultry he would be able to throw light on the vexed problems of evolution.

In the spring of 1900 de Vries, Correns and Tschermak independently discovered Mendel's paper, in each case publishing experimental confirmation of their own. Unfortunately de Vries did not mention Mendel's name in his first paper on the subject, though he used his terminology. This created an unfortunate impression which was very possibly undeserved. In the same year de Vries published results not only confirming Mendel but describing the remarkable phenomenon which he had observed in a series of experiments since 1886 on *Oenothera lamarckiana*, the evening primrose. This plant breeds nearly true, but produces a small proportion of abnormal forms, some of which in turn breed true. In 1900, then, Mendelism was launched. In December 1901 Bateson and Saunders published their first report to the Evolution Committee. They confirmed Mendel's laws on *Matthiola* and in particular on poultry. The hen was therefore the first animal for which Mendel's laws were found to hold. They were also proved on a number of other plants. In 1902 Cuénot extended Mendelism to mice, and in 1905 Farabee gave the first satisfactory account of Mendelism in man. Since that time Mendel's laws, with a certain amount of modification, have been found to be applicable wherever sexual reproduction occurs; and we now know the reason for their applicability is the remarkable uniformity in living organisms of the mechanism of meiosis, and the organization of the nucleus.

I will now point out a few of the landmarks in the first ten years of Mendelian research. In 1904 Bateson and Saunders and Punnett discovered the phenomenon now called linkage of genes. In 1903 Cuénot observed multiple allelomorphism in mice.

Starting from another angle in 1901 McClung discovered unequal chromosomes in the two sexes, and as early as 1902 Bateson and Saunders decided that sex was a Mendelian character, although it was not until 1910, as we shall see later on, that its connection with the sex chromosomes was made clear. Now all this time that Mendelism was developing there was a parallel development of the Biometric school, led in this country by Karl Pearson in London and Weldon in Oxford. Pearson's outlook was a positivistic one, and he covered an invaluable field of human heredity without any particular bias as to what theories his results might be expected to prove. He developed powerful mathematical methods for dealing with his material, but he was definitely rather weak on the experimental side. Bateson, on the other hand, though a poor mathematician, was a very great experimenter; and they did not always see eye to eye. In fact there was a violent polemic between them. For example, the biometricians published a series of pamphlets called "Questions of the day and of the fray" largely concerning Mendelian matters.

In 1905 an outsider might have said something like this: "I don't know whether Bateson or Pearson is right, but it is certain that one is wrong, and very badly wrong." He could even have taken sides on the basis of Oxford v. Cambridge. Curiously enough it turned out that both of them were right, and much righter than one could possibly have believed at the time. Bateson with the insight of a morphologist saw the atomic basis of heredity, and he was quite prepared not to worry about an occasional exception to the rules. Pearson and Weldon, on the other hand, severely criticized Bateson's work. They saw quite correctly that the early Mendelian theory was too crude and simple, and they gave particularly effective criticism to some of the early attempts to apply Mendelism to man. The present situation is, I think, as follows: in spite of the biometricians Mendelism is accepted by a vast majority of biologists, but if we want to discover whether a particular Mendelian hypothesis will explain a set of facts we are forced to use the mathematical criteria invented by Pearson. If we want the best examples showing Mendelian inheritance in man we have to turn to the *Treasury of Human Inheritance* started by Pearson, perhaps in the hope of disproving Mendelism. The synthesis between these two opposing schools has very largely been due to R. A. Fisher.

Now besides the Mendelian and biometric approaches there were three more or less independent lines of work in the early twentieth century. (1) Johannsen in Denmark elaborated the concept of the pure line and elevated it to the level of a scientific theory. (2) In 1907 Lutz, an American, made a very fundamental discovery. She found that whereas *Oenothera lamarckiana* has 14 chromosomes, its variety *gigas* has 28, while *lata* and *albida* have 15 each. This was the first example of a link up between cytology and genetics. Gates made some of these discoveries independently; they were published a few months later. (3) In 1909 Janssens gave the first satisfactory detailed account of meiosis. It had long been known that a diploid cell undergoing meiosis, that is, halving its chromosome number, divides into four individuals. That is why pollen-grains and spermatozoa are found in fours. Division into two would be sufficient to halve the chromosome number. Now Janssens was a Jesuit and he therefore had a teleological point of view in biology. He asked why there was an extra division. He said it could only be because it was intended that the four gametes formed by the division should all be different instead of being of two similar kinds. It is clear enough that the question could have been put from a Darwinian point of view. It could have been asked "why has natural selection brought it about that the division occurs into four and not into two"? It is, however, a fact that Janssens with his theological point of view was the first to put that question. He answered it by discovering what he called chiasmata, which are exchanges of material between paternal and maternal chromosomes and the cytological basis of the phenomenon of linkage.

A new epoch in Genetics began in 1910 with Morgan's work on *Drosophila melanogaster*. This is a small fly which has a generation in ten days; and you can get 400 from a single pair. It is clear that with such speed of breeding and such large numbers one can solve problems which are technically quite insoluble within a human lifetime with the organisms used by previous workers. At first Morgan and his colleagues got the normal type of Mendelian results—nothing particularly surprising. However, in 1913 things began to happen. His colleagues Bridges and Sturtevant made fundamental discoveries. Bridges, by a study of flies with extra chromosomes, proved that the sex-linked genes are actually carried by the X-chromosome, and Sturtevant made it highly probable that genes are arranged in a line along the chromosomes. He proved how from the strength of the linkage between the genes it was possible to make a map of the chromosome showing where each different gene was located. Bateson and many others held out for a considerable time against this interpretation. But Bateson came round to it, or most of the way round, before his death. Since that time *Drosophila* species have been classical objects for genetical study, especially for the relation between chromosome structure and genetical properties.

Another thing happened in 1913. Federley in Finland obtained in the second generation from crossing two species of the moth *Pygaera*, animals containing one set of chromosomes from one parent, as in normal hybrids, and two from the other. This was the first experimentally produced polyploid. In 1914 Gregory, a pupil of Bateson's at Cambridge, published a paper on the genetics of the tetraploid *Primula sinensis*, which has four sets of similar chromosomes and gives peculiar ratios among the offspring first explained by Muller. Finally, in 1917 Winge of Copenhagen called attention to the possibility of permanent and stable interspecific hybrids with two sets of chromosomes from each parent. It is interesting to note that such hybrids were first described in detail in 1926 independently by three sets of workers —Goodspeed and Clausen in California, Tschermak and Bleier in Austria, and Kihara and Ono in Japan. It is very striking how close together the results of various sets of workers in different countries were.

Questions of priority are almost trivial for that reason. Perhaps the last great discovery in genetics was that of Muller in 1927—that mutation could be enormously speeded up by X-rays. The unification of cytology and genetics was completed in 1933 by Painter who showed that genes could be exactly localized in the giant salivary gland chromosomes of *Drosophila*. Since then certain gene differences have been demonstrated to be due to visible changes in the chromosomes.

The main work of the last ten or twelve years in genetics has, I believe, been a work of unification. I will give you a few examples of this. Fisher showed that Mendelism was not merely compatible with the results gained by the Biometric school, but would explain a number of their peculiar features. Bell and Haldane discovered linkage in man and mapped out human chromosomes. Darlington, using the microscope, showed close correspondence between chiasmata and crossing-over observed genetically. Renner began the serious genetical analysis of *Oenothera*. Darlington showed how its peculiar cytology, and especially its habit—first discovered by Gates and Geerts—of forming rings of chromosomes at meiosis, explains the origin of the so-called mutants by crossing-over. The quantitative theory of sex-determination started by Goldschmidt and Bridges gradually crystallized out. In 1922 Vavilov enunciated the law of homologous variation. That is to say, he pointed out what had been vaguely known for a long time, that in related species similar mutants could be obtained and that they behaved in the same way genetically.

In predicting the existence of a number of new forms, comparative genetics was definitely launched. Comparative genetics shows a similar organization, in related species, of the genes whose co-operation determines the normal phenotypic condition in the species.

Practical applications of genetics are now being made on a very large scale. To give one simple example: about a million chicks are bred annually in England from sex-linked crosses, by breeders who report the numbers bred to the statistician. How many are bred by ordinary farms we do not know; but these million chicks owe their origin very largely to Prof. Punnett. Genetics has reacted on a very large number of other sciences, notably on biochemistry through the pioneer work of Garrod, who showed that a number of congenital abnormalities characterized by a deviation of some particular biochemical function follow Mendelian inheritance; for example, a failure of the oxidation of some particular amino-acid. The other most striking name in the borderline between genetics and biochemistry is that of Scott-Moncrieff, who has shown that in the formation of the anthocyanin pigments in plants a particular gene is responsible, perhaps for inserting oxygen at a particular point, for methylating a particular hydroxyl or for some other well-defined chemical process. Applications to immunology have also been important through the great work of Landsteiner, who has shown the existence in human beings of a series of genes determining the immunological properties of the red blood corpuscles. Applications to medicine have been very considerable, and in particular they are leading, I think, to a revival of the old idea of diathesis, a diathesis being a particular biochemical make-up, genetically determined, which may render a given individual particularly liable to a certain kind of disease. Some of the other applications are more controversial. Vavilov has applied genetics on a very large scale to prehistory. He claims to have determined the place of origin of a large number of different species of cultivated plants, and thus to have shown that cereal agriculture, for example, did not originate in the great river basins but in highlands in their neighbourhood. It is perhaps better to say nothing whatever with regard to the applications of genetics to anthropology. They are the subject of unscientific dogma in certain countries in central Europe and of embittered controversy in our own.

With regard to the question of eugenics, many workers believe that it is possible to apply genetical principles considerably to improve the existing human race. Others agree with the pronouncement of Bateson with regard to eugenics in 1925: "To real genetics", he wrote, "it is a serious, increasingly serious, nuisance, diverting attention to subordinate and ephemeral issues and giving a doubtful flavour to good materials." It is not for me to judge which of those two views is correct. But the most important application of all has, I think, been to evolutionary theory. At a fairly early stage it was proved that some of the differences regarded as interspecific are due to single genes. Perhaps the first clear case was brought forward by Lotsy in 1912 in *Antirrhinum*, but it is only one of a very considerable number. Other interspecific differences are plausibly accounted for by the interaction of a number of genes, but very often the differences between species turn out to be of a higher order, differences in the number or in the arrangement of the chromosomes; for example, in wheats it turns out that there are three distinct groups, with 14, 28 or 42 chromosomes, whose hybrids are more or less sterile. On the other hand in *Datura* the interspecific differences are largely due to the interchanges of segments between different chromosomes, the work on the subject being largely due to Blakeslee. It is by no means certain as yet whether all interspecific differences are of the same general character as inter-varietal ones. It is entirely certain that some of them are of the same character. Many very able geneticists, such as Goldschmidt, hold that some interspecific differences are *sui generis*. The most striking and complete modification of the Darwinian theory, I think, has been due to the recognition that an allopolyploid may become a new species. The landmark there was perhaps Müntzing's synthesis of *Galeopsis tetrahit*. *Galeopsis pubescens* and *Galeopsis speciosa* each contribute two sets of chromosomes, the synthetic form being produced with considerable difficulty by the doubling of chromosomes in the hybrid. It is sterile with the species from which it originated. It seems difficult to suppose that the species did not originate in the same way; and yet such a sudden origin is of course completely foreign to Darwinian ideas. It was Lotsy who probably laid the greatest stress on hybridization as a cause of the origin of species. Lotsy may have gone too far, like so many pioneers, but there can be no question that he pointed out an important truth, if he sometimes pointed it out a little too vigorously. Up to the end of his life Bateson was extremely sceptical as to the relevance of the fundamental discoveries which he and his colleagues had made for the problem of evolution. In 1922 he wrote as follows: "The production of an indubitably sterile hybrid from completely fertile parents which have arisen under critical observation from a single common origin is the event for which we wait." The difficulty, you see, is that the majority of varieties of species, even if, like the Pekinese and the Great Dane, they differ more than most species nearly related do, are yet fertile when mated together, whereas species commonly give no hybrids at all or else sterile ones. Bateson took the view in 1922 that such sterility had not been produced under critically controlled conditions. As a matter of fact it is almost certain that the event for which Bateson looked had already taken place in Gregory's cultures of *Primula sinensis* at Cambridge, in which tetraploids arose from diploids. These, on crossing back with the diploid, gave a triploid hybrid with three sets of chromosomes which was therefore sterile because of the very great difficulty of meiosis under such conditions. It was not until after Bateson's death that it became possible to produce tetraploids from diploids in large numbers and satisfy the criteria which he had laid down.

De Vries put forward the theory that species arise suddenly by mutation, and that was supported to a considerable extent by Willis' remarkable numerical studies of geographical distribution. Palaeontologists, on the other hand, found strong evidence for continuous change. It may well be that both are correct. Haldane has given rather unsatisfactory theoretical grounds for supposing that continuous variation would be most likely in numerous species, while a sudden origin of new species by mutation would be expected in the relatively rare ones.

In the last twenty years a considerable body of evolution theory based on Mendelism has arisen. The pioneer in mathematical theory was Norton, of Trinity College, whose work has been continued by Haldane, Fisher, and Wright; and there now exists a fairly highly developed mathematical theory of evolution. The gene rather than the individual is the most important unit considered. One is concerned with the spread of a given type of gene through a population. In so far as mathematical theory goes, there is a considerable formal analogy to statistical mechanics. One can represent a population as a point in n-dimensional space and study the trajectories of such a point. Whether this theory will prove an adequate explanation of evolution is much too early to state. It is important at any rate that it has made it necessary to define accurately a great many things which were only vaguely defined before. For example, the concept of fitness which Darwin used rather vaguely, and some of his successors much more vaguely, can now be defined numerically as a particular definite integral. The theory as developed is, I think, at bottom Darwinian, although it differs from Darwin in a very large number of details; it is interesting in that it offers one of the few possibilities of completely disproving Darwin's theory of natural selection. As soon as one can make a theory numerical, one is able to test it. Unfortunately the opportunities for testing are not very large, and fully controlled observations may require a good number of generations. The Lamarckian explanation has, I think, lost ground in recent years, since if not all, at any rate the vast majority, of alleged Lamarckian cases which have so far been thoroughly investigated have not been confirmed. It may very well be, on the other hand, that we are waiting for some entirely new principle in the study of evolution. Quite recently a number of workers have begun a study of the genetics of relatively large natural populations. The pioneer there was Tsetverikov, who showed that by inbreeding members of a natural wild population, which was apparently homogeneous, it was possible to disclose the existence beneath the surface of very large numbers of recessive genes. Turessen and Timoféeff-Ressovsky have investigated the adaptations of local races to such contitions as temperature and humidity, how far they remained when the species were transplanted into new environment, how they behaved on crossing, and so on. To many workers this was a relief in so far as it brought the genetical study of evolution back to earth after the somewhat speculative flights of Fisher, Haldane and Wright.

At present one may say that the mathematical theory of evolution is in a somewhat unfortunate position, too mathematical to interest most biologists, and not sufficiently mathematical to interest most mathematicians. Nevertheless, it is reasonable to suppose that in the next half century it will be developed into a respectable branch of Applied Mathematics.

In the forty years covered by my survey, genetics has come to occupy a somewhat central position in biology. Wherever individual differences are concerned, wherever you are studying not the rabbit, but this rabbit, you have got to take a genetical point of view. The future development of genetics depends, I believe, on its relations with other branches of biology. If it is kept to itself it may and probably will become sterile; a mere accumulation of formal facts about how various abnormalities are inherited. If, on the other hand, it is studied as part of the necessary equipment of all biologists it may, like anatomy when studied as a means to an end, become exceedingly fertile. It is not for me to say along which of these lines genetics will develop.